Exploring Central and Eastern Europe's Biotechnology Landscape

The International Library of Ethics, Law and Technology

VOLUME 9

For further volumes:
http://www.springer.com/series/7761

Peter T. Robbins · Farah Huzair
Editors

Exploring Central and Eastern Europe's Biotechnology Landscape

Springer

Editors
Peter T. Robbins
ESRC Innogen Centre
The Open University
Development Policy and Practice Group
Walton Hall
MK7 6AA, Milton Keynes
United Kingdom
p.t.robbins@open.ac.uk

Farah Huzair
ESRC Innogen Centre
The Open University
Development Policy and Practice Group
Walton Hall
MK7 6AA, Milton Keynes
United Kingdom
f.huzair@open.ac.uk

ISSN 1875-0044 e-ISSN 1875-0036
ISBN 978-90-481-9783-5 e-ISBN 978-90-481-9784-2
DOI 10.1007/978-90-481-9784-2
Springer Dordrecht Heidelberg London New York

Library of Congress Control Number: 2011934140

Printed on acid-free paper

Springer is part of Springer Science+Business Media (www.springer.com)

To Vaneeta for her ongoing love and support.

Acknowledgements

This book results from a workshop that brought together many excellent academic speakers, policy makers and industry representatives to discuss the current issues facing the life science industry in the region, and draw lessons from global case studies. We would like to thank AstraZeneca, Innogen, MEDON, OSI and PASOS for their generous funding of this event. We would also like to thank Adrian Ionescu (OSI) and our colleagues from Innogen at the Open University and Edinburgh for their support and helpful suggestions. Finally we owe a big debt of gratitude to Marlene Gordon for helping with the workshop organization and production of this manuscript. Of course, the usual caveats apply.

Contents

Contributors

Christien Enzing Technopolis Group, Herengracht 141, 1015 BH Amsterdam, The Netherlands, christien.enzing@technopolis-group.com

Erich Griessler Institute for Advanced Studies, Vienna, Austria, erich.griessler@ihs.ac.at

Farah Huzair Development Policy and Practice Group, ESRC Innogen Centre, The Open University, Milton Keynes, MK7 6AA, UK, f.huzair@open.ac.uk

Gyula Kasza Corvinus University, Budapest, Hungary, gyula.kasza@uni-corvinus.hu

Atina Krajewska University of Exeter, Exeter, UK, a.krajewska@exeter.ac.uk

Teresa Kulawik Södertörn University, Stockholm, Sweden, teresa.kulawik@sh.se

Zoltán Lakner Corvinus University, Budapest, Hungary, zoltan.lakner@uni-corvinus.hu

Aivita Putnina Centre for Bioethics and Biosafety, University of Latvia, Riga, Latvia, aivita.putnina@lu.lv; putnina@policy.hu

Thomas Reiss Fraunhofer ISI, Karlsruhe 76139, Germany, Thomas.Reiss@isi.fraunhofer.de

Laurence Reynolds Lancaster University, Lancaster, UK, l.reynolds@lancaster.ac.uk

Peter T. Robbins Development Policy and Practice Group, ESRC Innogen Centre, The Open University, Walton Hall, MK7 6AA, Milton Keynes, UK, p.t.robbins@open.ac.uk

Jacqueline Senker SPRU, University of Sussex, Brighton BN1 9QE, UK, J.M.Senker@sussex.ac.uk

Bronislaw Szerszynski Lancaster University, Lancaster, UK, bron@lancaster.ac.uk

Chapter 1
Human Life Science and Agricultural Biotechnology in Transition: An Introduction

Farah Huzair and Peter T. Robbins

1.1 Introduction

Human and plant biotechnology is progressing at a rapid pace in many countries across Europe. The development of these technologies and implications for society are highly controversial. There is an increasing awareness that in order to regulate effectively and form policy around these technologies, there is a requirement for public engagement, debate and new institutional capacities. Capabilities for the development of biotechnology have survived the process of transition in many former Soviet economies. The question of whether progress in the Central and Eastern European (CEE) economies is coping with the challenge of engaging society, whether they face the same legislative, administrative and ethical dilemmas is a compelling one.

European enlargement, accession and the changing structure of governance in the EU means that at this point it is crucial to understand how member states are able to engage with EU institutions in the debate on biotechnology and its future in Europe. This period represents a unique point in time when these dynamics can be investigated. And this book uniquely pulls together studies at the member state and European levels to answer these questions directly.

It may be argued that many countries, and not only the former transition economies, encounter difficulty in deciding how to incorporate the new technologies into their societies. This is particularly the case for less developed countries and societies that might be described as 'laggards' in technological development or public consultation. This book therefore offers useful and timely lessons and examples.

F. Huzair (✉)
Development Policy and Practice Group, ESRC Innogen Centre, The Open University, Milton Keynes, MK7 6AA, UK
e-mail: f.huzair@open.ac.uk

P.T. Robbins, F. Huzair (eds.), *Exploring Central and Eastern Europe's Biotechnology Landscape*, The International Library of Ethics, Law and Technology 9,
DOI 10.1007/978-90-481-9784-2_1, © Springer Science+Business Media B.V. 2011

1.2 Scientific Capability Under the Soviet Regime

Under the Soviet Regime, central planning controlled nearly every aspect of the economy including the education system, research, innovation and product development. The cornerstone of basic science research was usually the Academies of Sciences. Balazs (1997) describes the Academy 'model' as being developed in Russia and then rolled out to all the countries of the Soviet Bloc. Soviet politics influenced the organization of the academy by the focus on particular scientific fields. The natural and technical sciences, particularly those associated with military capability became dominant in most academies (1997). The institutes of the academy received basic funding from the central budget and this was dependent on the number of employees. And so the only incentive was to grow since an increased size meant more funding (1997).

Various studies in the 1970s report that rapid expansion throughout the USSR since 1928 resulted in an R&D sector comparable in size to the US. Progress was seen to be remarkable: 7600 innovations by 55 large firms in Poland during the period 1973–1978. However, if it is realized that 20% of these were failures, 80–90% were small innovations and minor in-house improvements with only 2% being classified as major structural innovations (Gomulka, 1986), it becomes apparent that the figures are much less impressive.

The science systems, culture and the academies, showed variation across the CEE Bloc despite the homogenisation concomitant with the Soviet model. Balazs (1997) argues that no two countries in the Bloc were more dissimilar than Russia and Hungary, Hungary being much smaller with less of a military industry and being the advanced in terms of market experience. One role played by Hungarian research institutions was to channel Western technology to the Eastern European market. They imitated, copied and reinvented Western products such as computer production. Biology was a relatively neglected scientific discipline across the Soviet Bloc not only because other disciplines were more closely related to military efforts but also because biological science programmes were more expensive to run in terms of equipment (Senker 2007).

T.R.Lysenko was director of Soviet biology firstly under Stalin in 1927 and then later under Khrushchev. Lysenko rejected Mendelian genetics in favour of the hybridization theories of another Russian horticulturalist Ivan Vladimirovich Michurin. Backed by unorthodox experimental and even fraudulent research he gained favour with Stalin due to his personal and ideological profile, won all manner of prizes and was promoted as an example of how practical solutions may triumph over theoretical research. His ideas gained momentum and created its own political movement 'Lysenkoism', promoted by the Soviet propaganda machine. It was also communist party policy to promote members of the proletariat into leadership positions rapidly. Born to a peasant family, Lysenko fit the bill. This explains why Soviet leaders generally welcomed his promises of increased agricultural yields through the idiosyncratic techniques he promoted. When the discovery of the double helix was being talked about with great excitement in the West in 1961, it was nearly impossible to publish anything regarding this in the USSR. Lysenko was finally dismissed in 1964 and was blamed for setting Soviet progress in biology back by years.

Despite these setbacks in the fields of biology and biotechnology, capabilities were maintained in some countries due to relative independence or the presence of key individuals. Those particularly noted for state of the art programs in biotechnology include Hungary and the Czech Republic (see Chapter 8, this volume).

1.3 National Science Systems During and After Transition

Market transitions throughout the Eastern Bloc occurred largely in the 1990s. It was of course accompanied by economic hardship due to the harsh austerity measures imposed by governments that attempted to control inflation and currency value fluctuations. In the midst of the economic crisis research and development were understandably not seen as priorities for the government. Research that had almost solely been undertaken by the public sector suffered from a shrinking allocation of funds, including in biotechnology. Fortunately capabilities did survive despite the ongoing problems of brain drain, lack of funding and a substantial loss of related industry and industrial connections. Biotechnological and other research activities in the former Soviet Bloc reflect a wide range of interests and sponsors. Private sector involvement in development varies from country to country. Likewise the structures of partnership, co-operation and collaboration between public, private and academia are also variable and dependent on technology regulation, the inherited institutional structure and government science and technology policy.

It has been argued that in an attempt to reverse the extremely interventionalist role of the state during the socialist regime, leaders and policymakers withdrew state control from many policy areas resulting in a missed opportunity to promote development during the transition (Von Tunzelman 2005). As such, some argue that there may have been a policy vacuum in biotechnology during critical points in the last few years, which impeded development in CEE states causing them to lag behind the rest of the EU.

However, research indicates that though the region during the transition has suffered from limited funding and institutional robustness, coupled with an absence of definitive government direction, this has not seriously affected capabilities for biotechnology development (see Chapter 8, this volume). The period has however brought the challenge of public involvement to CEE governments. In the past, Eastern European countries rarely had the resources, the expertise or the inclination to undertake costly participative democratic exercises.

In many former Eastern Bloc countries, the system of science and technology has developed in a very isolated way. Traditional innovation models are based on the experience of the West and do not reflect the dynamics of Soviet science systems, which were different in several respects. Under the communist model, research and learning were separated. Universities thus became merely ‘teaching factories’.

Specialised government research institutions, including the Academies of Sciences, undertook research. There was limited scope for the interaction of science with society. The scientific community developed neither the capability nor culture of communicating their work to a wider audience. Innovations and product development were delivered as uncompetitive inputs to a state owned industrial or military sector.

Years of isolating science from society have created a particular institutional environment. This is the case not only for science practitioners, but also for the policy-making institutions governing them. Acha and Balazs (1999) suggest that after many years of socialist rule, institutions in CEE have a resistance to and weak capacity for change. This is the result of an in-built inertia, itself a result of learning patterns that are both local and path dependent. Because new learning is built on previously established knowledge bases, it is virtually blind to other learning trajectories. New experiences are therefore interpreted according to established conceptual models, as is the case for policy makers in Central and Eastern Europe. From this it may be theorised that the adjustment of institutions will result in locally or nationally specific characteristics rather than imported models (Acha and Balazs 1999).

Policy-making institutions may face further barriers including the reluctance of public actors to share responsibility and accountability, not only with each other but also with private or societal actors (Teisman and Edelenbos 2004). There may also be a lack of transparency and openness in policy-making institutions, a remnant of the former command economy. Transparency, particularly with reference to dialogue between actors and stakeholders, is necessary if consistent policy outcomes are to be achieved (McQuaid 2005).

While innovation in the biotechnology field should usually be accompanied by dynamic adjustments at the institutional and social levels (Tzotzos 2000), as a result, social change can be difficult to achieve over short periods of time, and certainly cannot be expected to keep pace with physical and technological change. Green et al. (1999) affirms that there can be match and mis-match between new generic technologies and institutional arrangements that accompany take off. 'Boom' periods occur where there is mutual reinforcement between technological and institutional changes.

Janike et al. argue that the successful implementation of policy does not occur through the sole use of a single policy instrument but rather depends on attaining institutional capacity and the flexibility to use a combination of multiple tools over time. Institutional factors which limit capacity include 'political and legal structures and the rules and norms that produce a framework for interaction. In this area, participation, integration, decentralisation and the capacity for strategic planning are seen as particularly important' (Murphy 2001). Again the question that follows is whether transition economies have sufficient institutional capacity and motivation to achieve this goal.

Although significant change occurred during the transition to a market economy, it is difficult to tell to what extent transformations have occurred in the science sector. For example is there a willingness and acceptance by the science community to communicate and interact with new actors in new arenas? And secondly we may ask whether there has been change in society and the extent to which different publics have the capacity and have been enabled to engage with new knowledge, new sources of knowledge and create areas in which new technologies can be debated. Two factors demand these changes; the new geo-political situation in which Eastern European countries find themselves, and the developments of the biotechnologies themselves.

1.4 Geo-Political Transformations and European Accession

The eight 'vanguard' countries: the Czech Republic, Estonia, Hungary, Latvia, Lithuania, Poland, Slovakia and Slovenia, aimed to join the EU in 2004. The process of accession is dictated by the Copenhagen criteria, which states that national policies must align with the *Acquis Communautaire*, a massive body of legislation. Accession is therefore a bureaucratically complex task, though the formation of a national governance regime in biotechnology in CEE is challenging for a number of theoretical reasons.

Jehlicka and Tickle (2004) predicted that CEE countries will become passively compliant with EU regulatory and governance requirements and their national perspectives will become eclipsed by EU hegemony. They observe that generally the new accession countries have high stress sensitivity to the external environment, but low influence capacity due to the undemocratic nature of the harmonisation process as laid out in the Copenhagen criteria and a generally weak tradition in areas such as environmental policy. They draw on Thorhallsson's argument (2000) that integration behaviour is the result of administration size with small states having insufficient capacity to address all negotiations owing to a lack of staff, expertise and other resources.

Of significant concern was the theorised 'race to the bottom', in other words the competitive adoption of the lowest possible standards in terms of bio-safety and regulation. This fear has not materialised so far. Hungary, for example adopted particularly stringent national legislative forms of the *Acquis Communautaire* relating to agricultural crop biotechnology, seeing a niche for itself in the organic European market. Negotiation with the European Commission over national laws on co-existence of genetically modified crops and the de-facto moratorium banning commercialisation is still ongoing.

Yoder (2003) observes that in the 1990s the approach adopted by the EU with respect to biotechnology regulation shifted from 'government' to 'governance', during which time there was also a shift in emphasis towards multi-level governance and sub-national units of authority became more important. Governance, as opposed to government, can be defined as 'a complex process involving the interaction of multiple stakeholders often with different definitions of 'the problem' in numerous forms at different political levels' (Murphy and Chataway 2005). Bache and Flinders (2004) discuss the adoption of multi-level governance in the following way: 'the multi-level governance concept thus contained both vertical and horizontal dimensions. 'Multi-level' referred to the increased interdependence of governments operating at different territorial levels while 'governance' signalled the growing interdependence between governments and non-governmental actors at various territorial levels.' (Murphy and Chataway 2005)

This ideal stands in contrast to observations of policy discourse from authors such as Beck (1992) whose conception of 'risk society' may be judged as appropriate in the context of emerging science. Beck describes a tendency towards 'sub-politics' where political decisions are displaced to other non-political realms. These include 'the obscured worlds of laboratories, scientific councils etc'. Hajer (1995) similarly disqualifies the role of the layman as the understanding of science becomes increasingly technical.

The actual situation in CEE may tend towards Beck's notion of expert government due to the lack of the capacity of societies and publics to engage in debates around biotechnology. The normal protagonists of debate and key actors in civil society such as NGOs will not be present to the same degree as in the rest of Europe. While Carmin and Vandeveer (2004) speaking on a general level, report the increasing politicisation amongst the environmental and civil society actors and the aid made available by the EU and international sources to allow these groups to become further professionalised, Von Homeyer (2004) suggests that the position of NGOs in CEE is weak compared to those in longstanding EU member states. The degree of centralisation is evident as a factor that distinguishes the CEE former socialist style of policy making. This may limit integration due to the EU's requirement for the decentralised multi-level approach, particularly in environmental policy making but also perhaps in biotechnology.

In terms of scientific capacity, effects brought about by accession have been marginal. Scientists in Eastern Europe have for some time been members of wider global communities and networks. European science funding was at first preferentially given to candidate countries, but decreased quickly to competitive levels upon full membership of the European Union. With national funding much reduced due to economic difficulties, scientists in Eastern Europe increasingly look to European sources of funding such as the Framework Programs, and in doing so, have adjusted their research priorities and goals to European objectives (see Chapter 8, this volume).

1.5 Technological Shifts

We argued that lessons provided by the examples in this book would be useful to a wider range of countries than just those emerging from economic transition. This is due to the challenging nature of the new biotechnologies themselves. Tait (2007) describes new biotechnologies such as GM crops as being 'path-breaking' in nature. A path breaking technology is defined as one that challenges the innovation strategy, the market, or the regulatory system. The forthcoming chapters describe in detail how various technologies have indeed challenged markets and regulation at both the national and EU levels. With a few examples we might review here why that might be the case and how different aspects of different technologies offer challenges.

The controversies surrounding the GM debate have been well documented. In particular we might note the work comparing EU and US regulatory positions by Levidow, Murphy and Carr (2007). Of particular interest here are the characteristics of the technology which create controversy in Europe and which may cause difficulties for the regulatory authorities in transition countries.

GM crop technology highlights the problem of harmonisation in the EU. GM crop technology is similar to certain medical technologies such as pharmaceuticals, in that it can be realised in the form of a tradable good, capable of crossing borders, and therefore requires a harmonisation of certain regulations (e.g. product safety, labelling etc.) to enable trade to take place. However, it is unique in that it can also be manifested in the form of a living biological organism. Arguments exist which frame GM crops in terms of their possibility to cross-pollinate to weedy relatives, to 'contaminate' organic food production systems and as a plant species, 'disrupt natural ecosystems'. Framed in this way, GM crop technology becomes not a containable tradable commodity, but a technology with an inherent environmental risk, the use of which must be debated amongst the communities and publics concerned. Regulation under this regime must come from governance rather than government. In addition, governance must be at the EU as well as national levels.

GM crop technology creates a regulatory problem in that it may fall within the governance domain of more than one discipline or ministry. Competent authority may reside for example with the ministry for the environment or the ministry of agriculture. The multi-faceted nature of the technology means that many parties will claim involvement in regulatory decision making; those interested in trade, in bio-safety, environmental politics, science and industry, and so on.

Human embryonic stem cell (HESC) research is on the other hand, a much newer technology, whose proponents believe will have wide therapeutic impact, including treatment and alleviation of suffering of those with spinal cord injuries, burns, multiple sclerosis, and Parkinson's disease, as well as revolutionising tissue engineering and regenerative medicine. Debate on its governance and use is still at a relatively early stage in Europe. The motivation behind its production is fundamentally different to GM crops; it is not perceived as commercial product but a medical tool. It has however proven to be equally as controversial and the debate is just as polarised. There are similar issues around the different framings of the technology and arguments over definitions and meaning by different societal actors. HESCs are derived from an early stage human embryo and have the ability to develop into almost any kind of tissue, however the creation of stem cell lines involves the destruction of the embryo. At the heart of the controversy is how people view the status of the human embryo. Is the embryo a human life or does it represent the potential for life? Do embryos have a high or low moral status? Some in the pro-life movement believe that human life begins when a sperm fertilises an egg, and Roman Catholic doctrine for example holds that the deliberate destruction of an embryo is unacceptable; as such in strongly Catholic countries, like Poland (see Chapters 4 and 5, this volume) or Ireland, these definitional politics are shaped by framings of the high moral status of the unborn. Fink (2008) looked at a number of different types of embryo research in European countries and found a positive correlation between the proportion of

Catholics and restrictions on embryo research. Other contexts generate restrictions as well; in Germany rejection of Nazism has shaped a policy context that advocates uncompromising respect for human dignity (Barnes and Dupré 2008).

Within each national context cultural values are drawn upon to champion what is seen to be in the country's economic, scientific or ethical interests, in what Salter (2007) calls a 'moral economy' of the embryo, which is a field 'where values may be traded and cultural disputes routinized, though not necessarily resolved' (p. 270). While the components of this moral economy constantly shift, due to changes in the national and international policy context, it provides the basis for negotiation and compromise over divisive issues, such as source, date of creation, age, and research purpose of embryos. This has most notably been played out in the negotiations around the European Union's Sixth Framework Research Programme (see also Chapter 3, this volume).

A number of countries, including Latvia (see Chapter 6, this volume), Estonia, Iceland and the UK are compiling genetic databases involving the collection of personal information and samples from donors, with the stated intention of tracking genetic and environmental factors in disease in order to improve population health. This raises a rather different set of social and ethical questions to HESC research, which are around the obligations of researchers and rights of research participants to confidentiality, access to research outcomes, commercial exploitation and benefit sharing. This unfolds in the context of powerful economic and political actors, such as states, pharmaceutical and insurance companies, for whom access to genetic data allows for increased surveillance of populations, with potentially negative outcomes for civil rights, and financial liability.

Underpinning the concept of informed consent in this context is the idea that the donor is acting altruistically in providing their personal information and DNA as a gift so that future generations, rather than the donor him or herself will benefit. This notion is problematic if such a gift is used for financial gain, through the promotion of third party commercialisation and private property rights (Haddow et al., 2007). Donors are disrespected further, some argue, in that consent arrangements evade the moral obligation of informing research participants of health information that is clinically significant (Greely 2007). Finally, the framing of donors as 'research participants', when no participatory role is available to them in determining how their 'gifts' might be used and by whom, heightens ambiguity and may ultimately endanger public trust (see Chapter 6, this volume; Greely 2007, Tutton 2007).

1.6 Contributors to This Volume

The contributors to this volume have engaged with these issues using a number of different theoretical and methodological approaches. All to some extent examine how these issues are constructed and contested by key actors, government officials, industrialists, journalists, patient and other stakeholder groups, looking at what opportunities are enabled, and which are foreclosed by governments and publics. In other words, they flesh out biotechnological governance dynamics. Senker et al., (see Chapter 2, this volume) provide an overview of biotechnology development in the region between 2002 and 2005, looking at factors that are supportive of science and technology systems. They find that while development is occurring, particularly in terms of the acquisition of cutting edge tools and techniques from other countries, a focus on sectors like pharmaceuticals, where there is little supportive infrastructure is problematic, more successful are countries that are able to link science and technology innovation with areas of national economic strength. Erich Greissler asks the fundamental question, how participatory are EU institutions? His analysis at EU level using policy evaluation and interviews reveals the mechanics of the decision making processes, the opportunities for engagement with member states and why the debate on HESC continues to be a concern held primarily by social elites. Kulawik also examines the politics of the human embryo but her analysis is at the national level. Using Poland as the case, she examines the interactions between different national actors attempting to reconstruct the policy process through a discourse analysis of grey literature, parliamentary discussions and speeches. Krajewska considers the Polish case deconstructing the tensions and ambiguities that arise between Catholic social thought and free market economy articulated in laws regarding medical genetics. Putnina examines issues faced by the introduction of human biotechnology at the national level in Latvia. Her chapter picks up on the issue of ambiguity looking at framings of organ transplant comparing them with the Human Genome Project in Latvia. She uses interviews with those engaged in the governance debate, patients, and others engaged in the technology, plus media text analysis to unpick the social construction of this debate. In the second part of this book our attention turns to agricultural biotechnologies. Kasza and Lakner use media text analysis and survey data to establish the levels of public acceptance of GM crops in Hungary. They examine a number of issues that emerge over the course of the investigation including complex political dynamics involving national and international representation of public interests and the presence of 'trojan horses'. Huzair also examines Hungarian agricultural biotechnology, but her focus is on the national system of science and technology innovation, and in particular the role of actors in shaping policy and innovation contexts conducive to the development of biotechnology research capacity in Hungary. Reynolds and Szerzynski, using the case of GMOs, take a theoretical approach to European space examining how the issue is framed in different political arenas, who is involved and which discourses are permitted, and how GMOs are understood by different actors.

1.7 Conclusion

This chapter in introducing the volume focuses on the unique geopolitical context of CEE countries, and the particular challenges faced in human and plant biotechnology governance and innovation. During the Soviet regime biology was neglected due to its expense, mismanagement and incompatibility with military goals, however there was uneven development across the Bloc, with countries like Hungary experiencing relatively more success due to management of technologies and connections from the West. While the 1990s was a period of economic hardship across the region alongside the challenges of EU accession, and the perception was that a policy vacuum existed that could be exploited unscrupulously, in fact capabilities for biotechnology development were affected little. The key question was how issues of social engagement, crucial to the governance agenda would be addressed in the region. The contributions to this volume show that the politics of biotechnology in terms of the roles of key agents shaping policy contexts is similar to many other national contexts, how they unfold can be mediated by national factors, but in all cases we see power dynamics between local, national and international interests.

These generate tensions around religion versus economic and scientific development in definitional politics of the unborn child in Poland, which has effects on legal framings of medical genetics. They also are revealed in the case of GMOs in Hungary where policymakers must mediate commitment to biotechnology innovation with public concern about GM foods. Countries that are most successful are those that are able to use international/national dynamics to their own national advantage; it makes no sense to invest in pharmaceutical biotechnology when there is a weak local market that is unable to support it (Chapter 2, this volume). That said this is difficult where biotechnologies are concomitant with hegemonic and homogenised economic and regulatory spaces that can threaten local sovereignty (Chapter 9, this volume). This suggests that leaders in the region, like those in many countries, are faced with the challenge of how to manage human and agricultural biotechnologies that are part of a global political and economic space, to best address needs at national and local levels; ways leaders answer these questions are often shaped by cultural politics.

References

Acha, V., and K., Balazs. 1999. Transitions in thinking: changing the mindsets of policy makers about innovation. *Technovation* 19 (6/7):345–354.

Andrews, R. 1997. Martin Janike and the 'Berlin School' of environmental policy analysis: an American perspective. In *Umweltpolitik und Staatsversagen*, eds. L. Mez and H. Weidner. Berlin: Rainer Bohn Verlag.

Bache, I., and M., Flinders. 2004. *'Multilevel governance', Oxford scholarship online*. Oxford: Oxford University Press. http://www.oxfordscholarship.com/oso/public/content/politicalscience/0199259259/toc.html.

Balazs, K. 1997. Is there any future for the Academies of Sciences. In *The technology of transition*, ed. D.A. Dyker, 161–183. Budapest: Central European University Press.

Barnes, B., and J., Dupré. 2008. *Genomes and what to make of them*. Chicago: University of Chicago Press.

Beck, U. 1992. *Risk society: towards a new modernity*. London: Sage.

Carmin, J., and S.D., Vandeveer. 2004. Enlarging EU environments: Central and Eastern Europe from transition to accession. In *EU enlargement and the environment*, eds. J. Carmin and S.D. Vandeveer, 3–24. London: Routledge.

Fink, S. 2008. Politics as usual or bringing religion back in? The influence of parties, institutions, economic interests, and religion on embryo research laws. *Comparative Political Studies* 41 (12):1631–1656.

Gomulka, S. 1986. *Growth innovation and reform in Eastern Europe.* Brighton: Wheatsheaf Books.
Greely, H.T. 2007. The uneasy ethical and legal underpinnings of large-scale genomic biobanks. *Annual Review of Genomics and Human Genetics* 8:343–364.
Green, K., et al. 1999. The construction of the techno-economic: networks vs. paradigms. *Research Policy* 28:777–792.
Haddow, G. et al. 2007. Tackling community concerns about commercialisation and genetic research: A modest interdisciplinary proposal. *Social Science and Medicine* 64(2):272–282.
Hajer, M.A. 1995. *The politics of environmental discourse: ecological modernisation and the policy process.* Oxford: Clarendon Press.
Jehlicka, P., and A., Tickle. 2004. Environmental implications of eastern enlargement: the end of progressiveeu policy? In *EU enlargement and the environment*, eds. J. Carmin and S.D. Vandeveer, 77–98. London: Routledge.
Levidow, L., J., Murphy, and S., Carr. 2007. Recasting substantial equivalence: transatlantic governance of GM food. *Science, Technology and Human Values* 32 (1): 26–64.
McQuaid, J. 2005. Developing an integrated approach to risk: the ILGRA network. In *New modes of governance,* eds. C. Lyall and J. Tait, 89–106. Chippenham: Ashgate Publishing Ltd.
Murphy, J. 2001. *Analysing environmental policy: capacities and discourses.* Research Paper No.22. Oxford Centre for the Environment, Ethics and Society.
Murphy, J., and J., Chataway. 2005. The challenge of policy integration from an international perspective: the case of GMO's. In *New modes of governance*, eds. C. Lyall and J. Tait, 159–176. Chippenham: Ashgate Publishing Ltd.
Salter, B. 2007. Bioethics, politics and the moral economy of human embryonic stem cell science: the case of the European Union's Sixth Framework Programme. *New Genetics and Society* 26 (3):269–288.
Senker, J. 2007. *Biotechnology policies and performance in Central and Eastern Europe.* Conference paper presentation, Innogen Biotechnology in CEE, Prague, October 2007.
Tait, J. 2007. *GM as a path breaking technology.* Paper Presentation, EGN conference, London, October 2007.
Teisman, G.R., and J., Edelenbos. 2004. Getting through the twilight zone: managing transitions through process based, horizontal and interactive governance. In *System innovation and the transition to sustainability*, eds. B. Elzen, F.W. Geels, and K. Green, 168–190. Cheltenham: Edward Elgar.
Thorhallsson, B. 2000. *The role of small states in the European Union.* Aldershot: Ashgate.
Tutton, R. 2007. Constructing participation in genetic databases: citizenship, governance, and ambivalence. *Science Technology, and Human Values* 32 (2):172–195.
Tzotzos, G.T. 2000. Industrial biotechnology: challenges and opportunities. In *Biotechnology in the developing world and countries in economic transition*, eds. G.T. Tzotzos and K.G. Skyrabin, 1–13. New York, NY: CABI publishing.
Von Homeyer, I. 2004. Differential effects of enlargement on EU environmental governance. *Environmental Politics* 13 (1):52–76.
Von Tunzelman, N. 2005. Network realignment, policy sustainability and EU accession. In *Science and technology lessons for CEE countries,* eds. A. Inzelt, G. Zaman, and S. Sandu, 15–39. Bucharest: Expert Publishing House.
Yoder, J.A. 2003. Decentralisation and regionalisation after communism: administrative and territorial reform in Poland and the Czech Republic. *Europe Asia Studies* 55(2):263–286.

Chapter 2
Biotechnology in Central and Eastern Europe: An Overview of Performance and Policy Systems

Jacqueline Senker, Christien Enzing, and Thomas Reiss

2.1 Introduction

Central and Eastern European (CEE) countries have experienced an economic transformation since 1989 but the weakness of the life sciences during the era of central planning suggests that biotechnology research in CEE countries could be at an earlier stage of development than in other member states of the European Union. This makes it important to assess progress in biotechnology in CEE countries, because biotechnology is one of today's key enabling technologies, and has become the driving force of dramatic changes in innovation processes in many sectors.

The weakness of the life sciences in CEE countries towards the end of the last century is shown by an analysis of their publications in the ISI database during the period 1992–97. It shows that post-Communist countries had a relatively homogeneous research profile with a similarly unbalanced and narrow disciplinary structure. Their internationally recognised research strengths focused around physics and chemistry, but life sciences were relatively neglected (Kozlowski et al. 1999). Two factors explain this neglect: firstly, research strengths were those linked to the military/industrial complex. Secondly, the system favoured basic and theory oriented disciplines that were less dependent on expensive equipment, but the life sciences demanded large-scale, costly research and experimental work.

This chapter provides an overview of recent biotechnology developments in ten CEE countries: 8 countries that joined the European Union in May 2004 (Hungary, Poland, the Czech Republic, the Slovak Republic, Slovenia, Estonia, Latvia, Lithuania), as well as two that joined in January 2007 (Bulgaria and Romania). They are a very heterogeneous set of countries, ranging in population from less than 1.35 million (Estonia) to 38.2 million (Poland), but they share the legacy of central planning under the Communist regime. The overview covers the period 2002–2005 and considers these countries' performance in biotechnology as well as their policies and funding for biotechnology research and commercialisation.

J. Senker (✉)
SPRU, University of Sussex, Brighton BN1 9QE, UK
e-mail: J.M.Senker@sussex.ac.uk

P.T. Robbins, F. Huzair (eds.), *Exploring Central and Eastern Europe's Biotechnology Landscape*, The International Library of Ethics, Law and Technology 9,
DOI 10.1007/978-90-481-9784-2_2,

The effectiveness of policy is explored by analysing the relationship between national policy approaches towards biotechnology and the performance of the ten national biotechnology innovation systems. The information was collected and analysed as part of the BioPolis research project (Enzing et al. 2007) which combined qualitative and quantitative methods to provide an in-depth overview of national policy instruments to foster biotechnology growth and scientific and commercialisation performance in 32 European countries.

The next section will outline the methodology used for the BioPolis study, and discuss some of the difficulties encountered in applying it to CEE countries, especially the difficulty of using S&T indicators to assess the performance of CEE countries. Section 2.3 will present the science and technology (S&T) indicators used to identify clusters of CEE countries with similar performance, including the imperfect S&T indicators that are available for various aspects of these countries' performance in biotechnology research and commercialisation. Section 2.4 will discuss the context for policy-making in the ten countries, in terms of the way in which their policy-making systems are organised and Section 2.4 will discuss the funding used to promote biotechnology and the instruments used to distribute that funding.

The chapter concludes by identifying policy characteristics that appear to either help or hinder biotechnology development and considers the areas of biotechnology research most likely to support the economic development of CEE countries.

2.2 Methodology

The BioPolis study employed both quantitative and qualitative methodologies.[1] It used a common definition of biotechnology,[2] so as to facilitate comparability between countries and of data collected with published statistics. The quantitative methodologies included the use of science and technology (S&T) indicators to assess performance in biotechnology. It is essential to be aware of the limitations of using these indicators to compare the performance of CEE countries with OECD countries. CEE countries have gone through an economic transformation since 1989, and have made considerable progress in harmonising their S&T statistics with OECD standards. Nonetheless, there are still serious difficulties in using publications to assess the knowledge base and patents to assess the technology generated in CEE countries. The first problem concerns the effect of the legacy of the past on the publishing behaviour of CEE scientists, even though this has changed since 1989. The former planned economies of CEE countries had a closed character and their scientists did not form part of the international science and R&D community (Radosevic and Auriol 1999). Publication of scientific results and the

[1] This section provides a brief outline of the methodology. Complete details are available in Enzing et al. (2007).

[2] The definition was based on the conceptual definition and list of technologies used by the OECD, but added a third part: a list of application areas.

international communication of science were rather limited; the results of research were produced as "grey literature" and not as papers in journals. The second problem concerns the bias towards English language journals in publication databases. Publication indicators may omit the achievements of countries whose scientists publish in national journals because the use of English language poses a barrier, but evidence about this is anecdotal only. A recent investigation of this question did not cover CEE countries (Porter et al. 2002).

The use of patent data as a measure of technological performance also has its problems. This data is generally used to measure the performance of developed OECD economies that are at the innovation frontier. Its relevance for CEE countries is limited by the fact that these latecomer economies are not at the global innovation frontier; they are involved in technological catching up through imitative learning. Latecomer economies may have very little visibility in patent data during the early phases of catching up although, over time, their learning activities may lead to the development of innovation capabilities, which will then become reflected in patent data (Radosevic and Kutlaca 1999). The analysis of patents was further aggravated by the specific method applied (Reiss et al. 2004). For reasons of better comparability and to select higher quality patents we used patent applications to the European Patent Office (EPO). However, CEE countries would probably patent initially at their national patent offices because EPO applications cost much more than national applications. They would only consider making EPO applications as their innovation capabilities increased, Accordingly CEE countries are probably underrepresented at the EPO. Despite these limitations S&T indicators provide some indication of the development of biotechnology to date in the ten CEE countries.

In addition to collecting and analysing S&T indicators to assess national performance in biotechnology, the BioPolis project prepared national case studies based on a common methodology that was outlined in a guidebook. The guidebook specified the definition of biotechnology to be used and all the information to be collected for each national case study. This included general background about each country, including support for science and technology, an overview of relevant actors in the national biotechnology innovation system, both funders and performers, and details of policies and associated instruments to promote biotechnology. The national case studies also collected information on all the policies and instruments employed during the period 2002–2005. This information was based on desk research and interviews with responsible policy makers, using a standardised questionnaire to collect information on all policy-directed instruments and the expenditure on each instrument. Governments can use a broad set of instruments to stimulate biotechnology, as biotechnology activities cover a large part of the innovation chain: from basic research to market demand. For instance a research programme is a policy instrument as it constitutes a framework of goals to be achieved and serves as a basis for defining and planning specific research projects. Other examples are programmes that encourage collaboration between academia and industry, industrial research grants, support for centres of excellence, support for commercialisation of research, support for start-ups, programmes encouraging mobility of researchers, etc. A policy instrument can be a funding mechanism, but also a set of rules, laid down in legislation, e.g. for intellectual property rights (IPR). BioPolis includes only policy instruments that implement policy through funding mechanisms, excluding tax measures.

Governments use policy-directed instruments to implement their policies; BioPolis differentiated between biotech-specific and generic policy instruments. *Biotech-specific policy instruments* focus on influencing developments in biotechnology in particular and the intention is described in a policy document. *Generic policy instruments,* targeted at science and technology in general, can also contribute to the development and commercialisation of biotechnology. BioPolis also collected funding data for non-policy directed funding of biotechnology. The main reason for this is that in some countries funding through policy instruments is a relatively small part of biotechnology funding and non-policy directed funding is the most important funding mechanism for biotechnology. Non-policy directed funding includes funding which is part of structural governmental support for scientific education, research and research infrastructure. This type of funding is mainly given through block grants to research institutes and the open-call system of research councils. Basic funding for universities is not included. Some of the CEE countries were unable to provide full information about the policies and related expenditure for promoting biotechnology research, and the content of this chapter must therefore be treated with great caution.

After completing the ten national case studies, the CEE countries were grouped into three clusters with similar performance in biotechnology, in order to explore whether shared policy-making or funding characteristics in each cluster could explain performance. This approach has two inherent problems. Firstly, policy activity is just one of several factors that determine the performance of a national innovation system. Other factors that affect the achievement of policy goals include national economic conditions, as well as the institutional, cultural and legal characteristics. The second problem relates to the time lag between introducing a policy and its outcome. The information we collected on current policies provides little guidance on current performance in CEE countries; it depends on previous policies. Our knowledge of previous policies in CEE countries was limited. We therefore compared the CEE countries with the other 22 countries in the study and took advantage of lessons learned from previous studies (Enzing et al. 1999, Reiss et al. 2003). More information about performance indicators is presented in the next section.

2.3 Performance in Biotechnology

This section presents indicators of biotechnology performance in CEE countries to provide an overview of their scientific and technological development over recent years. The limitations of using publications and patent data to assess the S&T performance of CEE countries, as discussed in the previous section, mean that these indicators should be treated with great caution. Furthermore, no comparisons with other European countries are provided, because S&T indicators are a poor tool for this purpose. Indeed, the historical and institutional conditions for developing biotechnology in CEE countries are so different to those in other European countries that comparisons between the two groups of countries would be misleading.[3] For this reason this section focuses on the comparison of performance between CEE countries. Special criteria devised by the BioPolis team were used to identify data on publications and patents related to biotechnology.

The main indicator used to assess scientific activity was publications data from the Science Citation Index. Publications data were adjusted to reflect national population (per million capita: pMC) to improve comparability between countries. Publications data were also used to identify clusters of countries with similar performance. It was decided that the growth in publications output over three periods, 1994–1996, 1998–2000 and 2002–2004, was the most appropriate way to identify these clusters, because the capacity to increase and sustain growth of publications over time indicates that countries are building the capacity to "catch up".

Figure 2.1 shows that all CEE countries are below the average publications output per million capita (pMC) of the EU-25. It also provides a basis to cluster countries with similar performance into the following three groups:

- Cluster 1: the Czech Republic, Estonia, Hungary and Slovenia are closing the gap with the EU-25.
- Cluster 2: Poland and Slovakia[4] are making progress.
- Cluster 3: Bulgaria, Latvia, Lithuania and Romania have weak performance.

These three clusters are used to present information about the ten countries in all subsequent figures and tables.

Figure 2.2 the share of biotechnology publications in all publications, shows the increasing significance of biotechnology over time in every country except Lithuania and Bulgaria.

The publications output of CEE countries was also analysed by biotechnology area over time and compared with the EU-25. As shown by Fig. 2.3, CEE countries had a slightly different pattern of specialisation than the EU-25 in the period 1994–1996, with a much lower proportion of publications in the health area and more in every other area; this was most marked in environmental and generic biotechnology.

However, the distribution between biotechnology areas for 2002–2004 shows that CEE countries have been converging toward the EU-25 and not retaining their early pattern of specialisation. There is one exception, food biotechnology, where the proportional increase in publications has been greater than in the EU25.

We also analysed participation per million capita of CEE countries in three thematic priorities of the EC's Sixth Framework Programme (FP6) that covered biotechnology: (1) Life sciences, genomics and biotechnology for health; (2) the bionanotechnology section of Nanotechnologies; and (3) Food quality and safety. Participation in these programmes indicates recognition of the competence of national scientists by those from other countries.

---

[3] Data on the performance of other European countries is provided in Enzing et al. (2007).

[4] Slovakia is in Cluster 2 because it failed to sustain and increase its early publications output over time.

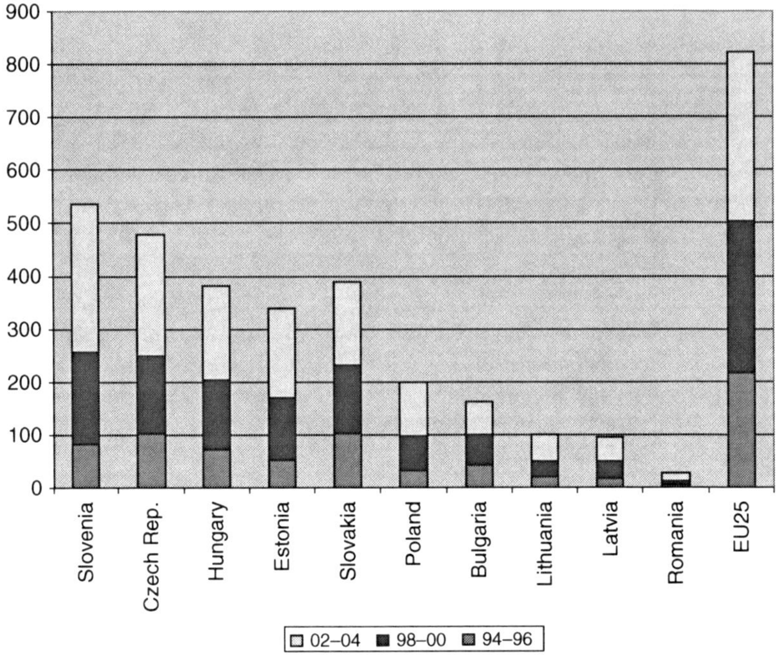

Fig. 2.1 Cumulative publications pMC in CEE Countries 1994–1996, 1998–2000 and 2002–2004. *Source*: BioPolis Research

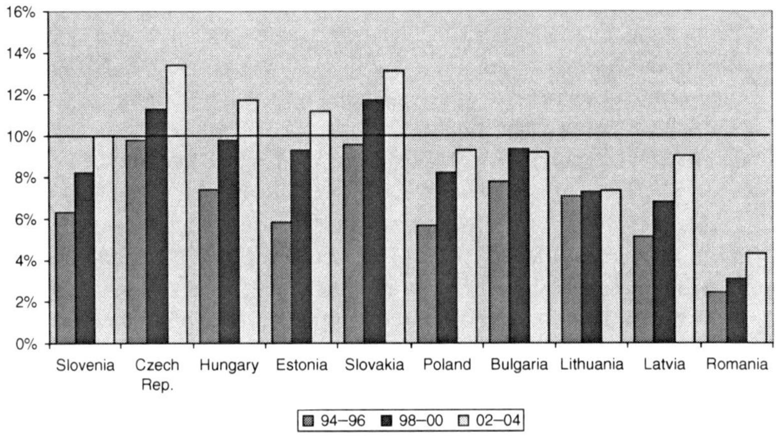

Fig. 2.2 Share of biotechnology publications in total publications (The *line* indicates the median for the ten countries in the period 2002–2004). *Source*: BioPolis Research

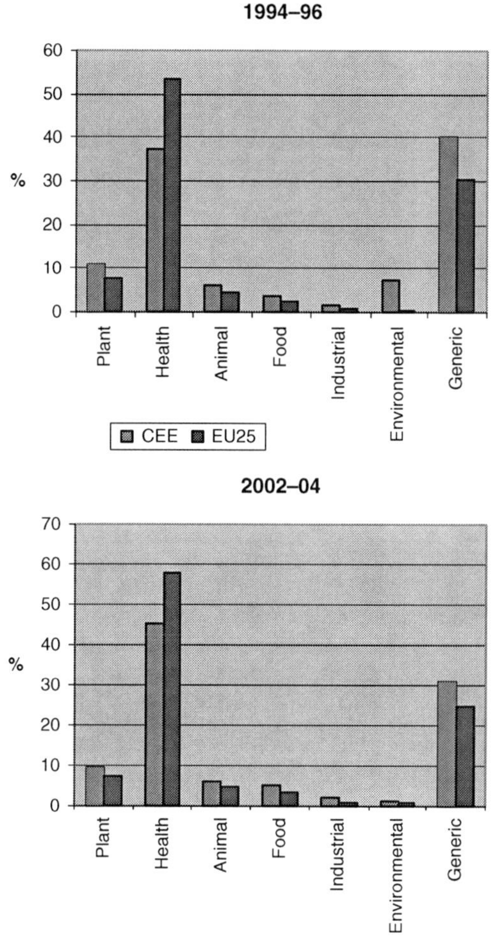

Fig. 2.3 Percentage of publications by biotechnology area, EU-25 and CEE countries, 1994–1996 and 2002–2004. *Source*: BioPolis Research

It also enables scientists in CEE countries to learn from other partners in the programme. Figure 2.4 shows the number of project teams in which each country participated, as well as the number of projects they coordinated.[5]

Only a few CEE countries have coordinated FP6 projects: Hungary (10), Poland (7), Slovakia (1) and Latvia (1). The Slovakian and Latvian coordinators were in the Food Quality and Safety thematic priority. So were five of Poland's coordinators and three from Hungary. Coordinations may reflect these countries' strengths in food and not in biotechnology.

[5]The number of coordinators has not been adjusted to national population (pMC) because absolute numbers were very small.

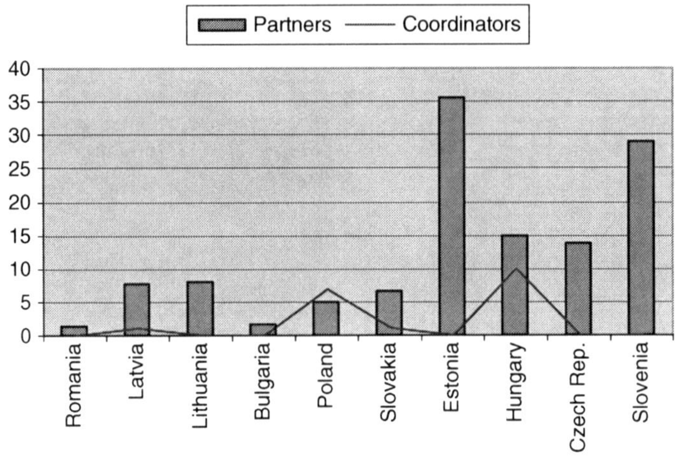

Fig. 2.4 Number of partners and number of coordinators in FP6 pMC. *Source*: BioPolis Research

The main indicator used to measure commercialisation performance is data on patent applications made to the European Patent Office (EPO). BioPolis also aimed to use data about biotechnology small and medium sized firms (SMEs) that was comparable between countries because it had been gathered using a common definition. However, the source[6] that covers many European countries does not yet include CEE countries. Other potential indicators for measuring commercialisation are the amount of venture capital invested in biotechnology firms and the number of initial public offerings (biotechnology firms floated on stock markets). Unfortunately, no such data was available for CEE countries, and this probably reflects the early stage of development of biotechnology in these countries. The limitations of technology indicators for measuring the commercialisation performance of CEE countries means that the data presented below, information on biotechnology patents and biotechnology companies for each country, must be treated with great caution. Figure 2.5 shows that several countries have increased their patenting activities over time.

Figure 2.6 indicates the number of biotechnology firms in each country, according to local estimates.[7] The data is not comparable because we do not know the definition used to decide which firms should be counted; therefore the figure should be regarded as a very rough indication only of countries where some degree of biotechnology commercialisation exists.

The next section reviews the institutional and cultural characteristics of each country as well as their policy-making systems, because these factors all contribute to performance.

[6]Ernst & Young reports.

[7]More than one estimate existed for some countries and sometimes the figures differed enormously. We have used the lower estimate in each case.

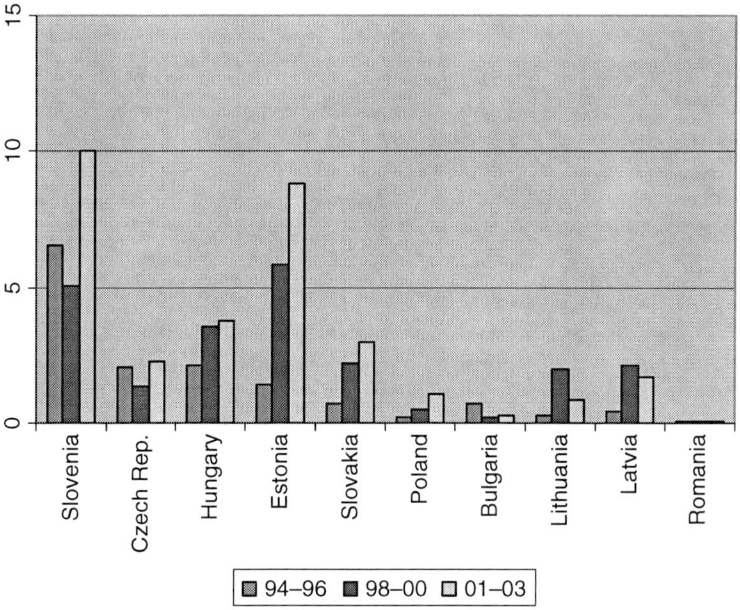

Fig. 2.5 Biotechnology patents per million capita (pMC). *Source*: BioPolis Research

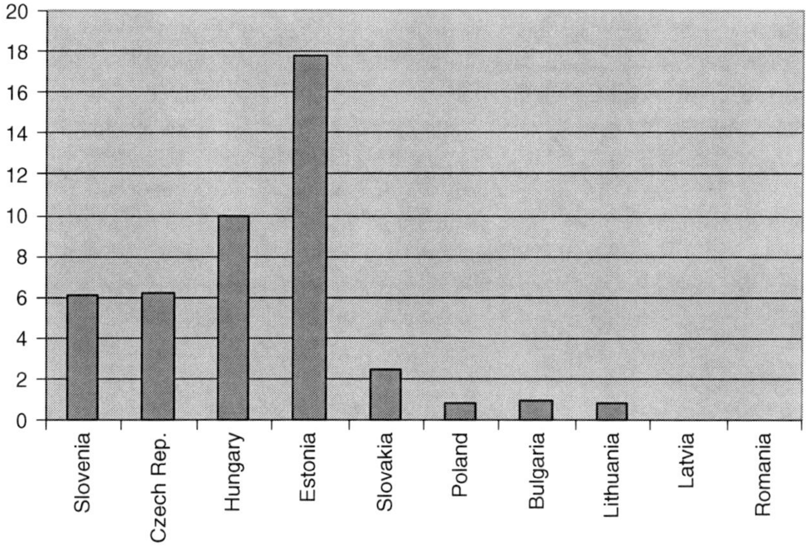

Fig. 2.6 Number of biotechnology companies per million capita (pMC). *Source*: BioPolis Research

2.4 CEE Countries' Policy-Making Systems

This section presents some basic data about the arrangements for policy-making in the ten CEE countries, in terms of some general economic characteristics, because this sets the strategic framework within which each country can define its R&D and innovation strategy. It will then present the policy-making systems of the three clusters of countries with similar performance in terms of (i) the main policy actors involved, coordination between the policy actors and the agencies involved in distributing research funds; and (ii) the researcher performers.

Before presenting these general characteristics, it is relevant to mention the shared heritage of the science systems of CEE countries that emerged from the regime of central planning during the Communist period. The science system was organised into three separate sectors, each with distinct functions. The National Academies of Science carried out basic research in institutes for the main disciplines and funding was allocated to these institutes, not to individuals or research groups. The heads of these institutes, Academicians, were responsible for making science policy to meet the plans set by their political masters and for coordination. Except for Poland and Hungary, where universities performed a significant amount of research (Radosevic and Auriol 1999), the higher education sector was devoted exclusively to education but, over time, it undertook some research, for instance work on research degrees. However, because the Academies and the universities competed for the same budget, there was unhealthy rivalry between them. The third sector, applied research and development, was carried out in industrial research institutes under specific ministries and was completely separate from the enterprises. There was little in-house industrial R&D (Balázs et al. 1995). The Czech Republic and Slovakia differ from this general pattern and over half of R&D was performed in the business sector (Radosevic and Auriol 1999). The economic crisis of the transition period, after 1989, led to a dramatic decline in resources for the research system (Balázs et al. 1995), and the system began to be restructured. The two main features of the restructuring are increased autonomy for scientists and the beginnings of competitive research funding (Kozlowski et al. 1999). Although the system is still in transition, Radosevic and Auriol (1999) foresee a "new division of labour" for national R&D systems in the long-term, and convergence with the R&D model of market-based economies in which business enterprises will perform R&D, and academic institutes and universities will be involved in basic research.

2.4.1 General Characteristics

The main characteristics of the ten countries in terms of their population, the intensity of R&D expenditure as shown by gross domestic expenditure on R&D (GERD) as a percentage of gross domestic product (GDP) and the existence of strong industrial sectors with the potential to exploit biotechnology are shown in Table 2.1. None of the countries reach the EU-25 average for GERD as a percentage

Table 2.1 General economic features of CEE countries

Country (population)	GERD/GDP 2004*(%)	Sectors relevant to biotech
Cluster 1		
Slovenia (1.99 M)	1.45	Pharmaceuticals
Czech Republic (10.2 M)	1.26	Chemicals, pharmaceuticals
Hungary (10 M)	0.88	Food, pharmaceuticals
Estonia (1.35 M)	0.88	Food & drink, wood processing
Cluster 2		
Slovakia (5.4 M)	0.51	Agriculture, food & drink
Poland (38.2 M)	0.56	Food
Cluster 3		
Lithuania (2.5 M)	0.76	Research materials, bio-pharmaceuticals, bioremediation
Latvia (2.3 M)	0.42	Food, wood processing
Bulgaria (7.7 M)	0.51	Brewing, dairy products, antibiotics
Romania (21 M)	0.39	Agriculture

Source: Eurostat website

of GDP, which was 1.86% in 2004; most are significantly below this figure. The industrial sectors relevant to biotechnology in the majority of the countries are traditional areas such as agriculture and the food and drink industry; several countries are involved in pharmaceuticals production.

The experience of the transition period made it difficult for some CEE countries to maintain or develop their biotechnology capabilities related to the pharmaceuticals sector. For instance, prior to the collapse of communism, Bulgaria produced antibiotics and had good research capabilities in fermentation technology and pharmaceutical biotechnology. As a consequence of the transition period, biotechnology-related capacities, infrastructures and resources deteriorated. Biotechnology products had been designed mainly for domestic use and for export to the Eastern Block countries. There was a collapse in demand from these markets, and Bulgaria was unable to enter new markets in the West because its products did not meet the requirements for handling genetically modified organisms.

2.4.2 Science and Technology Policy Actors

The configuration of the policy-making and research funding systems in many CEE countries has undergone many changes since the early 1990s, with frequent changes to Ministry responsibilities, the policy-making system, the agencies responsible for funding science, technology and innovation and their methods for funding research. Precise information on some of these issues is not fully available for all countries but rough judgments have been made from the facts that are accessible.

In most countries, responsibility for scientific research policy is vested in the Ministry for Education and Research (or similar) and responsibility for innovation policy in the Ministry of the Economy (or similar). The exceptions are Romania, where the Ministry of Education and Research is in charge of R&D and innovation policy.

Every country has an advisory body to support ministries in their policy-making and contribute to the coordination of science and innovation policy across government. Strong coordination of policy is supported by representation on advisory bodies of a wide range of actors with interests and knowledge relating to research and/or innovation policy. These actors include research funding agencies, the academic community and industry as well as ministers for agriculture, health or the environment, and national Academies of Science that have responsibility for science in their own institutes. In countries where there are separate advisory panels for science policy and innovation policy, coordination is supported by cross representation on each other's panels.

The academic community usually has some involvement in advising government on science policy. In most countries, this involvement is weak. There is less evidence of industrial involvement in innovation policy. Table 2.2 indicates how far each type of actor is involved in science, technology and innovation (STI) policy together with a judgment about the degree of policy coordination or fragmentation in each country. This judgment recognises that the need for coordination may be more important in large countries than in very small ones. The table is a qualitative assessment, based on the information contained in the national reports.

Another element affecting coordination is the extent to which research funds are allocated by research councils through a competitive, peer-reviewed process or in the form of block grants to Institutes. Previous research (Enzing et al. 1999, Reiss et al. 2003) suggests that the former system allows *ex ante* coordination, before the implementation of strategic decisions. By contrast, the funding of research through the allocation of block grants gives autonomy to organisations over the research agenda, and coordination can only be carried out *ex post*. Moreover, competitive research funding by research councils is not only flexible, it appears "to be a more effective method to achieve higher scientific performance than direct control of funds by research institutions" (Reiss et al. 2003). Some CEE countries have now adopted or are moving to a competitive, peer-reviewed process but, in many others, a high proportion of research funding is still allocated as block grants to Institutes and/or universities. However, these funds may be allocated to Institutes dedicated to a specific area of research, e.g. molecular biology.

The agencies that fund research are normally separated from those that fund its commercialisation through support to applied research, technology development, industrial research grants, university-industry research collaboration and measures to encourage the creation of small firms. To the extent that information is available, Table 2.3 presents information about the funding agencies that exist in each country and the activities they support.

Table 2.2 Estimated influence of key players involved in policy-making for science, technology and innovation

Country	Ministries		Funding agencies	Academy of science	Industry	Academic	Coordi-nation	Comments
	1–3	**>3**						
Cluster 1								
Slovenia	••	–	–	○	•	•	••	Policy integration across government and with external actors
Czech Republic	–	••	○	○	–	•	•	Policy integration across government and with academic community. No industry participation
Hungary	–	••	–	○	○	○	••	Policy integration across government and with external actors
Estonia	••	–	–	○	○	•	••	Integrated STI policy
Cluster 2								
Slovakia	••	–	–	○	–	○	○	No industry participation. Weak links between S&T and innovation policy
Poland	–	••	–	–	–	○	○	No industry participation
Cluster 3								
Lithuania	••	–	–	–	–	•	○	No industry participation; weak links between Research and Economy Ministries
Latvia	••	–	–	–	–	○	○	No industry participation
Bulgaria	••	–	–	–	○	○	○	Weak links between science and innovation policies
Romania	–	••	–	•	–	○	○	No industry participation

Strong ••
Moderate •
Weak ○
No influence –
Source: BioPolis Research

Table 2.3 Organisations funding research and type of funding

Country	Organisation	Type of funding
Cluster 1		
Slovenia	Slovenian Research Agency	Competitive grants for scientific and applied research
	Slovenian Science Foundation	Grants for young researchers
	Ministry of Economy	Use of PSR research by industry, research cooperation and promoting foundation of new firms
	Government sources	Core funding for public research institutes
Czech Republic	Ministry of Education Youth & Sports	Funds research programs at universities
	Ministry of Industry & Trade	Grants for industrial research projects
	Grant Agency of the Czech Republic	Competitive grants for public and private sector research
	Academy of Sciences	Competitive basic research grants for researchers in its institutes (core funds from government)
	Various Ministries	Grants for public and private sector research
Hungary	KPI	Competitive grants for R&D and for innovation projects involving academic-industry collaboration; promoting public-private partnerships; promoting creation of high-tech firms and innovation by SMEs
	OTKA	Competitive grants for basic research in public sector
	Academy of Sciences	Block grants to institutes
	Bay Zoltan Foundation	Applied R&D in own institutes
	Various Ministries	Support research in sectoral institutes
Estonia	Estonian Science Foundation	Competitive grants for basic and applied research by universities and institutes
	Enterprise Estonia	Supports science/industry collaboration and Centres of Excellence in research
Cluster 2		
Slovakia	Academy of Sciences	Block grants to institutes
	Agency for Support of R&D	Competitive research grants (public and private)
	Scientific Grant Agency VEGA	Competitive grants for basic research (public)
	NADSME	Supports innovation by firms in regions
Poland	Ministry of Education & Science	Grants for response mode, commissioned and strategic research; block research funds for universities; block grants to Academy of Science Institutes; block grants to R&D Institutes

Table 2.3 (continued)

Country	Organisation	Type of funding
	PAED	Funds for SMEs to support innovation
	Foundation FIRE	Support for innovative start-ups
Cluster 3		
Lithuania	Lithuanian State Science and Studies Foundation	Individual and project grants
	Academy of Science	Block grants to Institutes
Latvia	Latvian Council of Sciences	Competitive project grants for public research organisations
Bulgaria	Nat. Fund for Scientific Research	Has moved to competitive principle for distributing grants; no further information
	National Innovation Fund	Market-oriented applied research projects; promotes links between research institutes, industry and SMEs
	Academy of Science and Centre of Agricultural Science	Block grants to Institutes
Romania	Min. Education & Research	No. information on how funds allocated
	Romanian Academy	Block grants to Institutes

Source: BioPolis Research

Table 2.4 shows the institutional actors performing biotechnology research in each country and the number of each type of institution. Some of these institutions concentrate completely on biotechnology; others conduct biotechnology research on specific topics, as part of a broader programme of research. An emerging trend (mainly in cluster 1 countries) is the use of research instruments that encourage networking between researchers in Institutes and universities, and/or with industry e.g. virtual centre of excellence for biotechnology in Slovakia, the Estonian Biocentre, the research centre programme of the Czech Republic and Hungary's innovation cluster programme.

2.5 Funding of Biotechnology

This section presents data about the funding of biotechnology research in CEE countries and the instruments used to distribute those funds.[8] Some countries were unable to provide full information about expenditure on biotechology research and therefore Figs. 2.7 and 2.8 should be treated with caution. Amounts for Bulgaria,

[8]There were some errors in the data on the funding of biotechnology in New Member States and Accession Countries in Section 8.3 of the BioPolis Final Report. The correct data are included in this chapter.

Table 2.4 Institutional actors performing biotechnology research

Country	Universities	Academy of Science Institutes	Other public research institutes/centres
Cluster 1			
Slovenia	2	–	6
Czech Republic	6	10	2
Hungary	√[a]	2	3
Estonia	3	–	6
Cluster 2			
Slovakia	3	7	–
Poland	21	10	–
Cluster 3			
Lithuania	2	2[b]	1
Latvia	–	–	6[c]
Bulgaria	5	6	16
Romania	3	1	6

√[a]no data available; [b]1 located at university; [c]located at 2 universities
Source: BioPolis Research

the Czech Republic and Romania are underestimates, as we were unable to get expenditure information for several national programmes. In addition, these figures do not reflect the research in universities or Academy of Science Institutes that is funded through block grants (see Table 2.3 above). The gaps in the data on funding of biotechnology in CEE countries mean that the following figures give an indication of *minimum* total expenditure by the ten countries only. Thus, our calculation of total expenditure on biotechnology research and commercialisation of 234.8 M EUR must be regarded as a rough estimate. It represents 6.6% of all expenditure on biotechnology research in the 32 countries covered by BioPolis. It is doubtful whether the percentage contribution would be substantially higher even if fuller data on biotechnology expenditure had been available.

Figure 2.7 shows that, in absolute terms, the countries in Cluster 1 are spending much more on biotechnology research than the other countries.

To provide a fairer comparison, Fig. 2.8 adjusts the data for each country in terms of its population and its purchasing power. It also shows the proportion of each country's budget which is policy directed, covering both generic and biotech-specific policy, as well as the non-policy directed funding. Figure 2.8 shows that Cluster 1 countries' funding for biotechnology research is also higher in relative terms than the expenditure of countries in other clusters. Figure 2.8 also shows that CEE countries spend a higher proportion of funds on policy directed generic instruments than the average for the EU-25, and very little on biotech-specific instruments. In addition, it indicates that spending on non-policy directed expenditure in most countries is below the EU-25 average. We believe that this may be a distortion, caused by the inability to collect information on expenditure for biotechnology research through block research grants for Institutes in some countries.

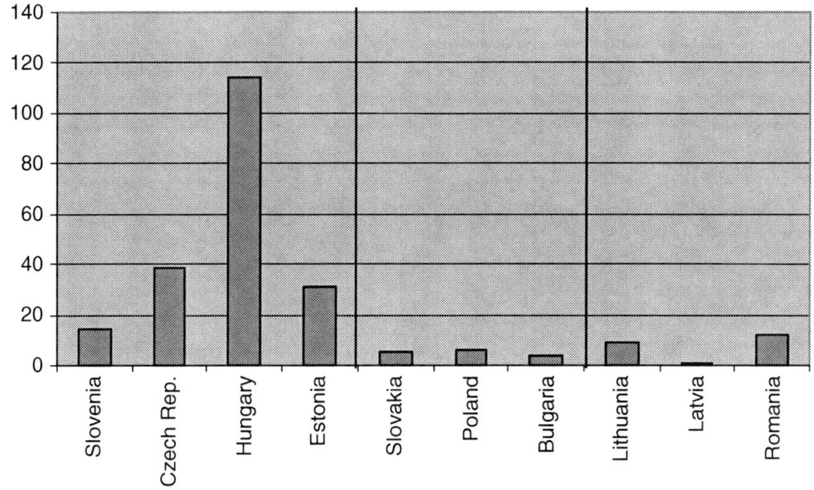

Fig. 2.7 Total budget for biotechnology (M EUR) 2002–2005. *Source*: BioPolis Research

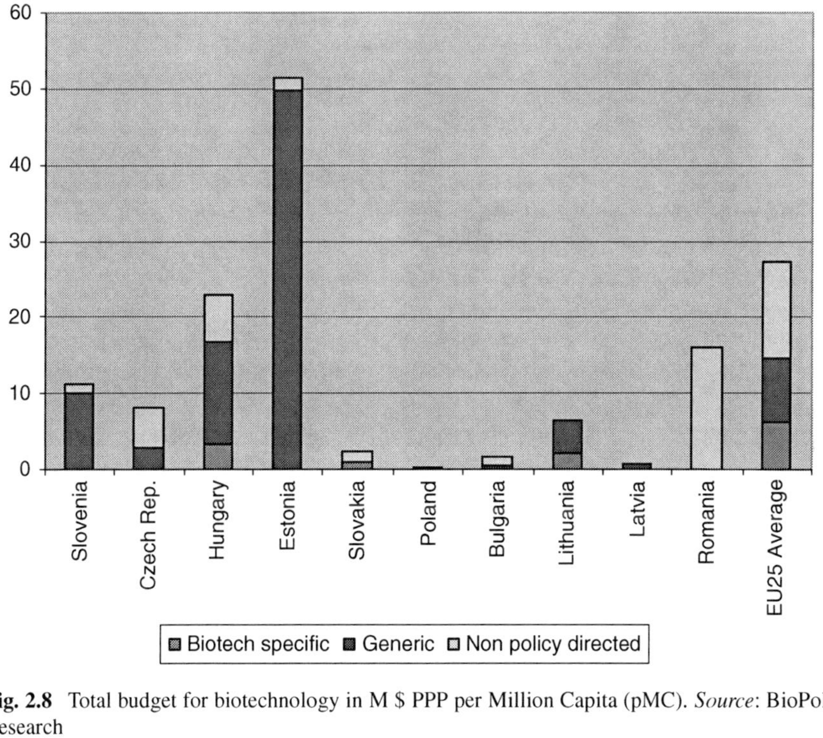

Fig. 2.8 Total budget for biotechnology in M $ PPP per Million Capita (pMC). *Source*: BioPolis Research

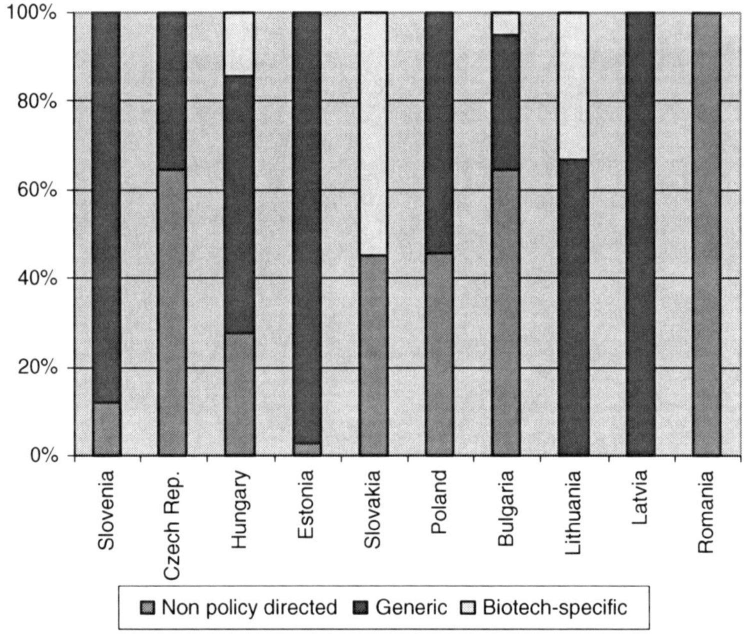

Fig. 2.9 Distribution of biotechnology funds by method. *Source*: BioPolis Research

We conclude this section by considering features of the policy-making system for biotechnology in the CEE countries. Figure 2.9 presents policy profiles for each country in terms of the proportion of non-policy directed and policy directed funds allocated to biotechnology. It shows the persistence of non-policy directed funding in many of these countries. Policy directed funding is broken down by the proportions allocated to support generic and biotech-specific research. The countries in Cluster 1, except for the Czech Republic, dedicate a significant proportion of funds to policy-directed instruments. Biotech-specific instruments are used only in a few countries.

Table 2.5 shows the number of instruments dedicated to policy-directed funding. Nine countries fund instruments to promote commercialisation, but only Bulgaria and Hungary attach this funding to biotech-specific programmes. Hungary and Slovenia are the only countries directing funds to "other" activities (activities to promote social acceptance of biotechnology, bio-safety or risk assessment). Overall, the countries in Cluster 1, the best performers, have a greater range and number of instruments than those in other clusters.

We next consider how far the identification of biotechnology as a policy priority has resulted in action to develop biotechnology. Biotechnology has been identified as a research priority in every country, except Romania and Slovakia. However, as shown by Table 2.5, only four countries have attempted to implement this priority by allocating funds to biotech-specific research programmes – Bulgaria, Hungary,

Table 2.5 Number of policy-directed research instruments by type

Country	Generic R&D	Biotech-specific R&D	Commercialisation Generic (B-S)	Others
Cluster 1				
Slovenia	3	–	2	2
Czech Republic	5	–	2	–
Hungary	2	3	3(2)	1
Estonia	5	–	6	–
Cluster 2				
Slovakia	–	3	1	–
Poland	2	–	–	–
Cluster 3				
Lithuania	2	2	1	–
Latvia	1	–	–	–
Bulgaria	2	1	2(1)	–
Romania	–	–	–	–

Source: BioPolis Research

Lithuania and Slovakia. Bulgaria and Hungary also allocate block grants to Research Institutes specialising in biotechnology.

Block grants for Research Institutes or university departments focusing on biotechnology are used as the main way to implement policy in two other countries that have biotechnology as a priority: Poland and Estonia. But neither biotech-specific programmes nor block grants for specialist public research organisations exist in three countries that regard biotechnology as a research priority: the Czech Republic, Latvia and Slovenia. Analysis of policy implementation is further confused by the fact that three biotech-specific research programmes exist in Slovakia, a country that has not identified biotechnology as a priority. These features suggest that the science and technology policy-making systems of most CEE countries are in an early process of development in terms of designing instruments that will allow the achievement of policy objectives.

Table 2.6 presents a summary of the strategies for biotechnology that were being developed after 2005. In general, there is little information on specific biotechnology policies or instruments, so information is also provided on general science and technology policy trends that may affect biotechnology.

2.6 Policy Characteristics Supporting Biotechnology Development

This chapter concludes with a discussion of the policy characteristics likely to support the development of biotechnology S&T in CEE countries. It is based on an analysis of the characteristics shared by the countries in each cluster, so as to distinguish the factors which might explain performance. The analysis identifies policy

Table 2.6 Future trends in biotechnology funding

Country	Future trends
Cluster 1	
Slovenia	No major changes to biotechnology are expected
Czech Republic	The National Innovation Policy 2005–2010 made no reference to biotechnology
Hungary	After major reforms in 2004, no major changes are expected in the short to medium-term. Elements to improve the current policy mix could be subsidies to reverse the brain drain, especially of industrial researchers, and various support measures for start-ups, including public seed and venture capital
Estonia	It is anticipated that new biotechnology specific initiatives will be launched
Cluster 2	
Slovakia	Priority areas in S&T for years 2006–2010 in development
Poland	The 2005 Act on Financing Science aims to concentrate expenditure on development projects that could be applied by SMEs, and to consolidate the R&D sector by promoting joint proposals. It also introduced the formation of instruments to support structural change, i.e. creation of science networks, consortiums and reorganization of the State Research Institutes.
Cluster 3	
Lithuania	The establishment of a biotechnology science park close to the main cluster of biotechnology research institutions and companies
Latvia	It is not clear whether specific biotechnology funding programmes are being developed.
Bulgaria	No major changes anticipated in the next few years. Insufficient funding for R&D will continue to be a problem. Consolidation of the large institutional research landscape will be attempted by setting thematic priorities.
Romania	Policy and instruments for biotechnology remain unclear after 2006

Source: BioPolis Research

approaches that could support the development of biotechnology in those countries currently making slow progress.

The limited information available for some CEE countries, and the problems with using S&T indicators to assess national performance, however, makes it difficult to draw strong conclusions about the policy characteristics supporting the development of biotechnology in these countries. Evidence from similar, earlier studies in other European countries provides further guidance (Enzing et al. 1999, Reiss et al. 2003).

It is clear that the history and traditions of each country can have a negative or positive impact on performance. CEE countries have had to overcome many barriers that hindered the development of their biotechnology capabilities, such as the neglect of the life sciences under communism and the former science system in which research was mainly carried out in Academy of Science institutes with universities involved only in education. For biotechnology, in particular, where strong links between public sector research and industry have been and remain crucial to the commercialisation of biotechnology, the historic separation between academic research and industry has been a major problem. Some, but not all CEE countries have made progress in solving these problems.

Previous history can have a positive impact on national performance in biotechnology. It appears that the existence of pharmaceutical companies in a country, as in Slovenia, the Czech Republic and Hungary (Cluster 1) can be beneficial, especially when the companies conduct research and are involved in science policy advisory bodies.

Government S&T policy is another factor that can explain the biotechnology performance of the three clusters of countries. However, clustering is based on publications performance in 2004, which resulted largely from the policy system in place around the turn of the century and not from the policy systems in the ten countries, which have been evolving rapidly over the past few years. However, we conjecture that Cluster 1 countries, the "catching up countries", have been quicker to adopt policy approaches recognised as supporting the development of biotechnology in other European countries, but usually absent from weak countries or those making slow progress (Clusters 2 and 3), and this may explain their performance. For instance, Cluster 1 countries spend more on biotechnology research in both absolute and relative terms than the other clusters. This cluster also has several funding agencies which support both public and industrial research, with the majority of funds allocated through instruments to implement policy. For instance Cluster 1 countries have instruments to support public/private research collaborations and networking, the valorisation of research, support for technology platforms, the creation of incubators and the formation of clusters. If this conjecture is correct then Cluster 2 and 3 countries can learn from the policy approaches adopted by the Cluster 1 countries:

- Allocating a significant amount of funds to research in general and to biotechnology research in particular;
- Creating policy instruments to implement research priorities;
- Ensuring policy coordination by having representation on policy advisory bodies of all the actors involved in biotechnology;
- Taking steps to move away from a system that allocates a high proportion of research funds in the form of block grants to Research Institutes and/or universities. It may be appropriate for these organisations to receive core research funds from government, but the quality and relevance of research is likely to be enhanced by greater use of competitive, peer reviewed research grants.

This chapter provides the first in-depth overview of the biotechnology policy making systems and policies in the CEE countries. These countries are mainly latecomers to the development and exploitation of biotechnology. They are correct to develop capability in this significant technology which has a rapidly expanding knowledge base. Without such capability they will lack the competence to absorb and utilise the knowledge that is being created in the rest of the world. For CEE countries there is a danger that in attempting to secure benefit from their investments in public biotechnology research these countries will focus on research related to the potentially high value-added area of applications of pharmaceutical biotechnology. Figure 2.3 above, which shows the increasing importance of research in the health biotechnology area, suggests that this is already occurring. However, this strategy has limited potential to support economic growth in countries which lack a strong pharmaceutical sector, as the competition in the rest of the world is so strong. In addition, pharmaceutical applications of biotechnology require capabilities in myriad new platform technologies. Building up an adequate knowledge base in even one of these sub-areas requires very large research teams and, particularly in small countries, would lead to the concentration of limited resources for biotechnology research in too few areas. CEE countries are more likely to gain economic benefits from their biotechnology research by identifying and supporting several areas of biotechnology research including research relevant to strong economic sectors within their countries, such as the food and drink or wood processing sectors.

Acknowledgements This chapter is based on a presentation prepared for The Promises and Challenges of the Life Sciences Industry in Central and Eastern Europe Workshop organised by INNOGEN, OSI and PASOS, and held in Prague, 18–19 October, 2007. The authors are grateful for support from the EC's FP6 Food Quality and Safety programme, which funded the BioPolis project through contract No. 514174. They are also grateful to their collaborators in the project: Annelieke van der Giessen and Sander van der Molen, TNO-Innovation Policy Group, Delft, The Netherlands; Ralf Lindner and Iciar Dominguez Lacasa, Fraunhofer Institute Systems and Innovation Research, Karlsruhe, Germany (Iciar Dominguez Lacasa is now at the Halle Institute for Economic Research, Germany); and Ismael Rafols, Pablo D'Este Cukierman and Janaina Costa, SPRU, University of Sussex, Brighton, United Kingdom. They appreciate both their work in preparing the national reports for the ten CEE countries on which this chapter is based as well as for their contributions to the ideas and the analysis presented in this chapter.

References

Balázs, K., W. Faulkner, and U. Schimank. 1995. Transformation of the research systems of post-communist Central and Eastern Europe: an introduction. *Social Studies of Science* 25(4): 613–632.

Enzing, C., et al. 2007. BioPolis. Inventory and analysis of national public policies that stimulate biotechnological research, its exploitation and Commercialisation by Industry in Europe in the Period 2002–2005. Final Report. Luxembourg: Office for Official Publications of the European Communities. Available online at:http://ec.europa.eu/research/biosociety/library/brochures_reports-BioPolis_en.htm

Enzing, C., et al. 1999. *Inventory of public biotechnology R & D programmes in Europe*. Luxembourg: Office for Official Publications of the European Communities.

Kozlowski, J., S. Radosevic, and D. Ircha. 1999. History matters: the inherited disciplinary structure of the post-communist science in countries of Central and Eastern Europe and its re-structuring. *Scientometrics* 45(1):137–166.

Porter, A.I., J.D. Roessner, X. Jin, and B.C. Newman. 2002. Measuring national 'emerging technology' capabilities. *Science and Public Policy* 29(3):189–200.

Radosevic, S., and L. Auriol. 1999. Patterns of restructuring in research, development and innovation activities in central and Eastern European countries: an analysis based on S&T indicators. *Research Policy* 28(4):351–376.

Radosevic, S., and D. Kutlaca (1999), Assessing the basis for 'catching up' of Eastern Europe: an analysis based on US Patent Data. STEEP Discussion Paper No.42, SPRU, University of Sussex, Brighton, http://www.sussex.ac.uk/Units/spru/publications/imprint/steepdps/42/steep42.html. Accessed 24 Oct 2007.

Reiss, T., S. Hinze, I. Dominguez Lacasa, V. Mangematin, C. Enzing, A. van der Giessen, S. Kern, J. Senker, J. Calvert, L. Nesta, and P. Patel. 2003. *Efficiency of innovation policies in high technology sectors in Europe (EPOHITE)*. Final Report. Luxembourg: Office for Official Publications of the European Communities.

Reiss, T., S. Hinze, and I. Dominguez Lacasa. 2004. Performance of European member states in biotechnology. *Science and Public Policy* 31(5):344–358.

Chapter 3
Citizen Participation in Controversial EU Research Policies? The Debate on Human Embryonic Stem Cell Research Within the 6th Framework Programme

Erich Griessler

> *Among the dilemmas frequently encountered by ethics, there is that born of the confrontation between freedom of research and* ***freedom to conduct business on the one hand, and the respect due to human life on the other*** *(EGE 2001: 11, emphasis added).*

> *Europe is currently at a crossroads: we need to actively develop responsible policies in a forward-looking and global perspective, or we will be confronted by policies shaped by others, in Europe and globally. The technology and its applications are developing rapidly – the Commission believes that Europe's policy is, therefore,* ***not whether but how to deal with the challenges posed by the new knowledge and its applications*** *(European Commission 2002a: 9, emphasis added).*

> *The Commission proposes a strategy that responds with* ***responsible, science-based, and people-centered policies on an ethical basis*** *(European Commission 2002a: 10, emphasis added).*

3.1 Introduction

Many social scientists agree that the EU institutions are confronted with a significant legitimacy deficit in biotechnology and life sciences (Abels 2002, Gottweis 2003, Salter and Jones 2002), a policy field in which they face severely diverging claims by science, industry and parts of civil society. This legitimacy deficit in the politics of new biotechnology also challenges traditional ways of policy-making in European institutions, particularly with regard to its way of expert involvement and reliance on technocratic networks. Thus, the European institutions are "seeking

E. Griessler (✉)
Institute for Advanced Studies, Vienna, Austria
e-mail: erich.griessler@ihs.ac.at

P.T. Robbins, F. Huzair (eds.), *Exploring Central and Eastern Europe's Biotechnology Landscape*, The International Library of Ethics, Law and Technology 9,
DOI 10.1007/978-90-481-9784-2_3, © Springer Science+Business Media B.V. 2011

new methods of engagement both with the expanding numbers of NGOs in the human genetics arena and the public at large" (Salter and Jones 2002: 326). This move towards participation, however, is not only fuelled by the "laudatory desire for greater citizen involvement in the governance of biotechnology" but also by the plain realization that consumers decide with their consumer decision on economic success or failure of biotechnology products (ibid). The "sea-change in the political culture of governance policy-making at EU level" (ibid.) is connected with learning from difficulties encountered by European institutions in their attempts to regulate genetically modified organisms (GMOs). Thus, the rhetoric of participation heavily influences the way in which the Commission talks about handling biotechnology. This is expressed, e.g., in strategic documents such as the "White Paper on governance" (Commission of the European Communities 2001a) or the paper "Towards a Strategic Vision of Life Science and Biotechnology" (Commission of the European Communities 2001b). The latter paper explicitly states, "transparency, accountability and participatory approaches in public policy-making need to be reinforced. These objectives coincide with those of the Commission's White Paper on European governance and will be pursued through the actions proposed therein" (European Commission 2002a: 20).

Gabriele Abels diagnoses the notion of participation as a "remarkable shift" in the Commission's position on "how to govern bio politics". For her, "'Participation' is the key word – yet reduced to a very limited concept (...). The proposed 'participatory' modes of governance aim at greater inclusiveness of social actors, i.e. experts and lay-people, stakeholders and citizens, the public and Autocrats in supranational policy-making and regulation. The underlying assumption is that the effectiveness and efficiency, i.e. the output side of policy-making, can be improved by strengthening the input-side and, in doing so, the legitimacy of EU policy will increase" (Abels 2002: 2). She notices three features of EU research and technology policy-making. Firstly, the system is heavily science driven and strongly relying on expert advice: "The Commission sets up and makes use of scientific advisory committees; it utilizes its organizational resources and experiences to decide who gets access to European policy networks and who doesn't. Policy- and decision-making on issues of science and technology is, above all, a 'politics of expertise'. Epistemic communities, that is networks of professionals with recognized expertise and an authoritative claim to policy relevant knowledge have easy – and above all – privileged access to European policymakers" (Abels 2002: 6). Secondly, Abels perceives a rise of bio-ethics, and thirdly, she observes an attempt to "take into account the social prerequisites of technological innovation" (ibid.). However, the notion of participation is in conflict with the "scientification of politics", i.e. the inflationary use of scientific expertise in public policy-making" (Abels 2002: 3).

Yet, how "participatory" are the European institutions when it actually comes to dealing with controversial issues of new biotechnology? In order to address this question, my paper will look at the way in which the Commission has involved "the public" in its decision-making process about whether human embryonic stem cell (HESC) research should be included in European research funding under the Sixth Framework Programme (FP 6).

EU regulatory power regarding "red" biotechnology, in contrast to "green" biotechnology, is rather limited since health policy – except regulation of pharmaceuticals – as well as the regulation of ethical issues of medicine and human reproduction are mainly national domains. The Commission seems to acknowledge the member states' responsibility in this area. At the same time, however, it seems to be inclined to foster "harmonization" for economic reasons. As Commissioner Busquin, for example, emphasized in his reaction to the UK vote on therapeutic cloning: "The Commission has no intention to legislate or harmonize in the field of ethics and respects this diversity of cultures and points of views in Europe." Nevertheless, in the same press release, he declared his intention for harmonization: "The realization of the European Research Area also entails that we look where and how we can foster an area of shared values in Europe. I have already taken initiatives to launch a dialogue between society and science at the European level and am committed to continue in this direction" (IP/00/15019 2000).

Embryo- and HESC research became a very controversial issue within and among the EU institutions in the context of research and technology policy in the Sixth Framework Programme (FP 6). The case is well documented in recent analyses using different approaches (Salter and Jones 2002, Capps 2005, Gottweis 2003; particularly Adenberger 2005, Pichler 2005). In my paper, I shall therefore focus on the question which participatory possibilities emerged in what way during the debate on funding; whether FP 6 funding should also target HESC research at all, and if so, under which conditions. I shall first sketch the role of the EU institutions, i.e., the Commission, the Council and Parliament, in decision-making. I shall then take a closer look at processes by which the Commission tried to support its decision-making and at the role which participation played in this. In the final part I shall present my conclusions.

3.2 Decision-Making

3.2.1 The European Commission

The Commission perceives biotechnology and life science – and accordingly HESC research – clearly within the context of the central Lisbon goal, "to become a leading knowledge-based economy" (European Council 2000). The Commission refers regularly to this goal and expresses its opinion that life sciences and biotechnology are key technologies leading to economic prosperity:

> Life sciences and biotechnology have entered a stage of exponential growth, opening up a vast potential to move economies in Europe and globally towards more sustainable development and improved quality of life. They are therefore of strategic importance in Europe's quest to become a leading knowledge-based economy. Europe cannot afford to miss the opportunity that these new sciences and technologies offer (European Commission 2001b: 3; see also European Commission 2002a: 8).

To give another example for this line of argument by the Commission: "[T]he life science revolution was born and is fed and nurtured by research. (. . .) There is an undisputed link between research, innovation, the competitiveness of industry and the generation of wealth and social prosperity" (European Commission 2001b: 11 ff.). However, the Commission is also concerned that "Europe's current performance in life sciences and biotechnology is not facilitating the achievement of that objective" (European Commission 2002a: 8).

Consequently the Commission supported stem cell research already during the Fifth Framework Programme (1998–2002) within the thematic programme "Quality of Life and Management of Living Resources". This decision was backed by an opinion formulated by the "European Group on Ethics and New Technologies" (EGE), an advisory body to the European institutions.

The Commission requested an opinion on embryo research in September 1998. In its opinion No. 12 of 23 November 1998 EGE confined embryo research to "experiments on embryos which are not intended for transfer to the uterus, and which do not survive" (EGE 1998: 78). EGE emphasizes that "the progress of knowledge of life sciences, which in itself has an ethical value, cannot, in any case, prevail over fundamental human rights and the respect which is due to all the members of the human family" (ibid.: 81). It states that "the human embryo (...) deserves legal protection", which "falls within the competence of national legislation". However, the Community authorities "should be concerned with ethical questions resulting from medical practices or research dealing with early human development". In that they should take into account "the moral and philosophical differences" between Member States, "the respect for different philosophical, moral and legal approaches and for diverse national culture" being "essential to the building of Europe" (ibid.). EGE states that "under the Community's Fifth Framework Programme, Community funding should not a priori exclude human embryo research (...) but that this funding should, nevertheless, only be granted under strict conditions" (ibid.: 82).

These are: "systematic ethical evaluation, at Community level, of protocols of research on human embryos presented for Community funding"; "priority should be given to the principle of the respect due to human life, as well as, respect regarding the consent of the women or couple concerned"; the project must comply with national regulations; where embryo research is permitted by national legislation, public as well as private research should be carried out "under strict public control" and "maximum transparency". Such transparency "should be a compulsory requirement of any proposal funded by the 5th Framework Programme, since it provides the best guarantee against major risks of arbitrary experimentation" (ibid.).

The opinion moreover emphasises the importance of enlarging public debate, which "is just getting underway". It also asks for additional Community money within FP 5 for global scientific and ethical evaluation of research projects involving human embryo research, the results of which should be made public. Finally, the EGE opinion desires that the Commission should create a system of information "regarding all ethical and legal aspects relative to life sciences, at both national and international level" (EGE 2001: 83).

Under FP 5, the Commission funded "15 research projects in the area of stem cell research and therapy with a total EC contribution of €27.4 million" (European Commission 2002b: 10).

In FP 6 (2003–2006), HESC research was mentioned under priority 1, Life Sciences, Genomics and Biotechnology for Human Health, as "focus on the development and testing of new preventive and therapeutic tools, such as somatic gene and cell therapies (in particular stem cell therapies)" (Commission of the European Communities 2003b: 4). In preparation for the new Framework Programme, the Commission came to the conclusion that another EGE opinion should back the decision whether to include HESC research in funding (Gottweis 2003: 17). In short, the EGE opinion took the position that HESC research should in principle be allowed, but with certain restrictions (EGE 2000a).

The EGE addresses "ethical issues raised by human stem cell research and use in the context of the European Union research policy and European Community public health competence" (ibid.: 14). It points out several "fundamental ethical principles at stake" such as "respect for human dignity", "individual autonomy (entailing the giving of informed consent, and respect for privacy and confidentiality of personal data)", the principles of "justice and beneficence", "freedom of research", "proportionality (including that research methods are necessary aims pursued and that no alternative more acceptable methods are available)". Moreover, EGE points out the "potential long-term consequences of stem cell research and use for individuals and society" (ibid.: 15). Again, EGE mentions "respect for different philosophical, moral or legal approaches and for diverse cultures" which is "implicit in the ethical dimension of building a democratic European society". In this connection, EGE refers to Article 22 of the Charter of Fundamental Rights and to Article 6 of the Amsterdam Treaty (ibid.).

EGE observes that embryo research is forbidden in some Member States, but allowed in others "for the purpose of treatment of infertility". It states that "*it is hard to see any specific argument which would prohibit extending the scope of such research* in order to develop new treatments to cure severe diseases or injuries" (ibid., emphasis in original). Following from that, EGE sees "no argument for excluding funding of this kind of research from the Framework Programme (. . .) if it complies with ethical and legal requirements as defined in this programme" (ibid.).

EGE states several requirements to be considered with respect to funding: Taking the UK Human Fertilization and Embryology Authority (HEFA) as an example, HESC research should be carried out in countries where it is allowed "*under strict public control by a centralized authority*" (ibid., emphasis in the original). Given the sensitivity of the "use of embryonic stem cells", authorization for both private and public research should be highly selective and transparent and be based on a case-by-case approach. EGE states that "the creation of embryos for the sole purpose of research raises serious concerns since it represents a further step in the instrumentalization of human life". It considers the creation of embryos "with gametes donated for the purpose of stem cell procurement" as "ethically unacceptable, when spare embryos represent a ready alternative source". Moreover it states that the "*creation of embryos by somatic cell nuclear transfer for research on stem cell therapy would be premature*" (emphasis in the original), since there are alternative sources from spare embryos, fetal tissues and adult stem cells.

EGE also appeals for "EU funding (that) should be devoted to testing the validity of recent discoveries about the potential of differentiation of adult stem cells" and states "a specific responsibility (at European Union level within the Framework Programme of research) to provide funding for stem cell research" (ibid.). This implies the establishment and provision of sufficient means for ethical ex-ante assessment and monitoring. Moreover, EGE "stresses the necessity *to ensure that the demand for spare embryos and oocytes does not increase the burden on women*" (ibid., emphasis in the original).

As in all previous EGE Opinions, the present one, too, was issued in consensus. However, the paper represented a compromise within the group: "This Opinion (. . .) was thus adopted by consensus . . . though some in the Group tended to oppose all human embryo research while others were more favorable to the development of 'therapeutic cloning'" (EGE 2001: 9).

This opinion, as Commissioner Busquin declared, served as basis for Commission policy: "In the preparation of future research programmes the Commission will base itself on the opinion of the European Ethics Group, especially on the opinion on the ethical aspects of human stem cell research delivered on 14 November 2000" (IP/00/15019 2000). Commissioner Busquin was guided in his decision by an "anticipated positive response by the majority of Member States, the positive opinion of the EGE, and because of the positive development in his native country, Belgium, where even the Catholic University of Leuven has been in favor of stem cell research under certain conditions" (Pichler 2005: 266). Subsequently, in 2001, equipped with the right of initiative in this matter, the Commission entered negotiations about FP6 with a permissive proposal regarding funding of HESC research.

3.2.2 The European Council

By contrast, political actors representing the member states within the Council held (and still hold) very different views regarding the use of HESCs for research purposes. National legislation has varied to a great extent across Europe (Commission of the European Communities 2003b; Capps 2005). The Council was split into a "permissive" and a "restrictive" faction. The permissive group included the UK, Sweden, Denmark, Finland, France, Belgium, Greece and the Netherlands, the restrictive one comprised Germany, Italy, Austria, Luxembourg, Portugal, Spain and Ireland (Pichler 2005: 267). This division roughly mirrors the group of Protestant and Catholic countries in Europe. Religion, however, should not be overestimated and was only one contributing factor in this debate. There are strong economic and strategic interests in HESC research (Salter and Jones 2002: 329, Romeo-Casabona 2002: 504, Salter 2005). Thus, different ethical standpoints as well as economic interests were major centrifugal forces in this conflict within the Council. However, since HESC research represented only a very small part within the framework programme, and since international economic competitiveness is a cardinal goal of EU policies, none of the players within the Council wanted to jeopardize the timely implementation of FP 6 as a whole.

A group of representatives from member states taking a restrictive position criticized the Commission's permissive approach in the Council and the Austrian representative even vetoed the Council's common position in June 2002. On 30 September 2002, the restrictive faction requested a moratorium to postpone actual funding of HESC research until 31 December 2003. The compromise stated that no funding for HESC research would be granted until 31 December 2003 except for isolated and banked human embryonic stem cells in culture. Moreover, it was agreed that the Commission would provide a report on HESC research, which was to serve as basis for an inter-institutional seminar in spring 2003 (see below). In addition, the Commission was to work out funding guidelines for HESC research.

3.2.3 The European Parliament

Salter and Jones perceive the European Parliament as the most active actor in the "expression of civil society interest" in biotechnology so far (Salter and Jones 2002: 334). The Parliament takes a more sceptical position towards HESC research than the Commission, which has been expressed in several resolutions concerning cloning and embryo research (see Salter and Jones 2002: 331, Romeo-Casabona 2002: 495). Salter and Jones explain this sensitivity with the MEP's dependency on their electorate:

> Members of European Parliament are naturally sensitive to the cultural response of their constituencies to human genetics developments. Unlike permanent officials, they are obliged to balance the pressures of the Brussels-based trans-national policy networks with the electoral consequences for themselves and their parties of failing to heed the often strongly held views of the citizens. As a consequence, they act as a conduit for the expression of a diverse range of ethical views on human genetics and find the achievement of a workable consensus in this field a less than straightforward matter (Salter and Jones 2002: 332; see also Adenberger 2005).

Like the Council, the European Parliament was split into a restrictive and a permissive faction. However, the permissive faction, consisting of Social Democrats, Liberals and a group of Conservatives from the European People's Party, finally prevailed in November 2003 in a vote of 298 in favor, 241 against, with 21 abstentions.

The provisions for funding of HESC research were: "The human embryos used for the procurement of stem cells must be supernumerary early stage (i.e. up to 14 days) human embryos (embryos genuinely created for the treatment of infertility so as to increase the success rate of in vitro fertilization (IVF) but no longer needed for that purpose and destined for destruction); such research may be funded provided that it is legally permitted in the Member State(s) where it will be conducted under the rules and strict supervision of the competent authorities" (European Parliament 2003: Amendment 10). Funding is also allowed for "research on embryo or fetal stem cells deriving from spontaneous or therapeutic abortion" (European Parliament 2003: Amendment 10; quoted in Pichler 2005: 266).

In April 2003 the Commission Staff Paper (Commission of the European Communities 2003b) finalized in December 2002 was presented and formed the basis of the previously mentioned inter-institutional seminar on HESC research held in Brussels in the same month. The Commission also worked out a proposal, based on Article 166(4) of the Treaty for guidelines "on the principles for deciding on possible Community funding of research projects involving in particular the use of human embryonic stem cells" (Commission of the European Communities 2003b: 12). This proposal was submitted to the Institutions in July 2003 and supported by the European Parliament (Commission of the European Communities 2003b: 17).

In November 2003, the Commission adopted procedural modalities for research activities involving banked or isolated human embryonic stem cells in culture to be funded under FP 6 (Commission of the European Communities 2004: 17).

At the end of the moratorium, the restrictive and the permissive factions within the Council did not agree on the funding condition for HESC research in the Specific Programmes within FP 6. In the absence of general guidelines, the Commission started to fund HESC research on a case-by-case basis after the expiration of the moratorium.

So far, only a small number of research projects on HESC have been funded. In its first call, the Commission funded 25 research projects in the amount of approximately €160 million, involving at least one component of stem cell research. More than 90% involved the use of adult human stem cells and only two use components of HESCs (from existing HESC lines). In 2005, in the 2nd call, the Commission expected to support 17 projects involving stem cell research, amounting to €110 million. Only one project would involve a component of HESC research (using existing lines; Commission of the European Communities 2005: 29).

3.3 How "Participatory" Was Decision-Making?

In the following section I will take a closer look at the role of public participation in the decision-making processes I described earlier.

3.3.1 Process of Decision-Making

Despite the rhetoric of participation and public dialogue the decision on HESC research was still prepared by a small number of officials within the Commission and finally taken by top politicians in the Council, the Commission and Parliament. In this process, formal powers and formal procedures as well as informal negotiations and bargaining at top level played a decisive role. A Commission official emphasized that in contrast to standard practice, according to which civil servants prepare decisions, actually top politicians took decisions on HESC research funding under FP6 at the last moment:

> [A] lot was really at high level, between the Ministers. Less than what we usually . . . Because usually it is true, everything is prepared. The Ministers go and they sign. And it is technical people who do their work. But this time it was not only the technical people who did the work. A lot was done in the discussions between the Ministers, between Commission. (. . .) It was very much, let's say, it was really a *political decision* (Interview 2002, emphasis in the original).

Such a decision-making process, not surprisingly, was the exact opposite of the idea of openness vis-à-vis the public and public participation.

3.3.2 How Does the Commission Handle the Challenging Problem of HESC?

However, how does a bureaucratic organization like the European Commission in principle approach and tackle a problem such as the controversial issue of HESC research funding with European research money? I would like to distinguish several types of activities in this respect, i.e., information gathering and report writing, negotiating, incorporating the new topic into administrative routines and obtaining some backing by experts.

One way of dealing with such a problem was to follow administrative routines and to compile reports on international HESC research regulation (European Commission 2001b, 2004) and actual research (Commission of the European Communities 2003b). As a Commission civil servant said, the inter-institutional seminar held in Brussels in April 2003 provided an opportunity to discuss these reports with other actors from relevant EU authorities.

> [T]he report is a basis for the discussion. So the report is supposed to give them information about (. . .): Where are we with the science? What are the legal situations? What are the main ethical issues? So it should give them the background to try to formulate their opinion on this issue, on as I said whether we should be allowed to fund research involving the use of spare embryos. And also what should be then the guiding principles for such research? If it should be allowed and in general (. . .), what should be our policy? (. . .) What are our needs if we want to try to promote this area of research? At EU level or under the Sixth Framework Programme? How can we contribute, I mean to keep Europe competitive in this area? So let's say also I hope they even have a more also general discussion on the whole issue. And let's say the European policy, because there is also competitiveness in World and or, or, or industries (Interview 2002).

However, this seminar addressed mainly members of the EU institutions and some outside experts. The role of the public was restricted to a spectator on the Internet.[1]

[1] http://ec.europa.eu/research/conferences/2003/bioethics/index_en.html (download 2007-01-21).

Another way of coping with this topic was to incorporate it in the Commission's already existing administrative routines of dealing with research proposals and to add an ethical review of proposals on research funding:

> A specific ethics review has been implemented for proposals dealing with specific and sensitive issues such as the use of banked or isolated human embryonic stem cells in culture, human fetal tissue or cells, non-human primates, animal cloning, human beings, genetic information etc. The recommendations from the ethical review are taken into account in the negotiation of the projects (Commission of the European Communities 2004: 19).

Yet another way of approaching the problems posed by HESC research was to add expertise to the decision-making process by inviting experts or creating new bodies, thus making political room for experts on ethics as well as on biotechnology and life sciences in general, and particularly on stem cell research. An example for the approach to invite experts was the conference "Stem Cell Research at European Level" held in September 2001 in Brussels, which involved 12 of the 15 co-coordinators of research projects on stem cell research at that time funded by the Commission (European Commission 2001a). Examples for the creation of new bodies are the European Group on Ethics in Science and New Technologies (EGE) as well as the European Group on Life Science (EGLS).[2]

3.3.2.1 The European Group on Ethics in Science and New Technologies (EGE)

Salter and Jones identify the establishment of the European Group on Ethics in Science and New Technologies (EGE) in 1998 as one of the two main developments, which are important for the establishment of bio-ethics in the EU. This group started in 1991 already, as "Group of Advisers to the European Commission on the Ethical Implications of Biotechnology" (EGE 2001: 2).

[2]The Forum of Presidents of National Ethics Councils is yet another example how the Commission tries to tackle the difficult issue of ethics by taking a top-level expert approach. The Forum consists of the chairpersons and secretaries of the National Ethics Councils and is to promote harmonisation and benchmarking by open co-ordination: "It is an independent informal platform for exchange of information, experience and best practices on issues of common interest in the field of ethics and science. The NEC Forum follows the method of 'open co-ordination' and its meetings are always hosted by one of the National Ethics Councils. The Commission (Directorate-General for Research) reimburses the travel and subsistence costs of one representative per National Ethics Council. The NEC hosts the meeting while DG Research provides the secretariat. The President of EGE and the President of the COMETH (Council of Europe) Bureau are invited to the meetings. The Forum network is developing an important role in exchanging good practices between Member States" (http://ec.europa.eu/research/science-society/page_en.cfm?id=3161) The Forum was created in the context of the EU's Science and Society Programme (Commission of the European Communities 2004: 20). Between 2003 and 2006 it met eight times in different Member States and at these meetings it is often received by top politicians. Stem cell research was an issue several times (current therapeutic possibilities of adult and embryonic stem cells 5th Meeting: 2). Supported by the Commission, the Forum might contribute to the establishment of a particular epistemic community and expert network on bio-ethics in Europe.

The task of EGE is to "examine ethical questions arising from science and new technologies and on this basis to issue Opinions" (EGE 2001: 158). It "is an independent, pluralist and multidisciplinary body" (ibid.: 3) and issues its Opinions on request of "the Commission, the Parliament, the Council, or on its own initiative" (ibid.: 159). Its Members are appointed by the European Commission "for their specific skills" and come from different disciplines and professions. They "deliberate freely and in total independence in accordance with its rules of procedure. They also communicate the Group's opinions as they see fit, naturally ensuring the Commission receive them first" (ibid.: 4 ff.). The EGE sees its role:

> Ethics must (. . .) help the Community authorities, which are responsible for regulating the market, to take better account of the aspirations of the public in the various aspects of their lives: as consumers, workers, parents, patients etc. With this in mind, the Group intended to reattach to European ethics the principles which are not always directly associated with it; the idea being to create a relationship of trust between science and society (ibid.: 12).

Salter and Jones perceive the EGE as an important and self-confident policy broker with good links to national bio-ethics committees and international communities such as the Council of Europe's Steering Committee on Bioethics and UNESCO's International Bioethics Committee. They conclude, "in the fluid politics of the EU it is a player to be taken seriously" (Salter and Jones 2002: 336). The Commission attributes great importance to EGE and repeatedly emphasizes its increasing significance:

> The Commission welcomes the key role played by the European Group of Ethics in Science and New Technologies since its creation in the early 1990s and proposes (. . .) to enhance its role and to reinforce the networking with and between national ethical bodies (European Commission 2002a: 20).

The EGE's high status is also underlined by the fact that it handed over its Opinion on HESC research to the President-in-office of the European Union (IP/00/1293). The Commission stresses that it will enhance the role of EGE by closer collaboration with the Commission services and increasing the exchanges with other institutions, in particular Parliament (Commission of the European Communities 2004: 20). The EGE also strengthens its collaboration with National Ethics Committees.

Salter and Jones are, however, sceptical about the role of bio-ethics in decision-making:

> Bioethics presents itself as both expert and as having a hotline to the needs of civil society through its impartial consideration on moral concerns. By linking its claim to legitimacy to a quasi-representative function in this way, bioethics may be able to resolve, or at least to ameliorate the effects of, regulatory conflicts whilst this occurs within the relative cerebral confines of the EU policy community and its immediate network environs. However, whether its legitimacy will survive prolonged public exposure to a media-driven issue in human genetics is unknown and untested (Salter and Jones 2002: 338).

Though claiming, as already said, to voice the "aspiration of the public" (EGE 2001: 12) § 25 of its "Rules of Procedure" states explicitly "the deliberations of the

Group are confidential" (EGE 2001: 160). EGE also concedes that it has a problem with public involvement in its Opinions:

> Generally speaking, the Group has endeavored to be as open as possible in its proceedings. It is true that the Group's deliberations on its Opinion are not public, but otherwise the EGE Round Tables are open to representatives of interest groups, MEPs interested in any given matter, delegates from other international bodies (EGE 2001: 5).

Thus, participatory elements in the EGE setting are restricted to 1-day round tables, in which established stakeholders are consulted but not able to decide. Opinion No. 15 on HESC research in the context of FP 6 refers to such a round table organized in June 2000 in Brussels "with members of the European Parliament, jurists, philosophers, scientists, representatives of industries, of religions, of patients' associations, and of international organizations (Council of Europe, UNESCO, WHO)" (EGE 2000a: 2). In addition, there were two hearings of representatives of experts and one hearing of representatives of religions on 8 September 2000 (ibid.). Looking at its list of participants, this round table had 82 participants: 12 members of EGE, three participants from the EGE secretariat, seven invited speakers, 14 representatives of the Commission, four members of the European Parliament, three participants from the Economic and Social Committee, six representatives of international organizations (WHO, Council of Europe, UNESCO, European Patent Office, OECD), nine representatives of media, four from religions and 20 experts (EGE 2000b: 223–227). From the participation point of view, the round table clearly favored policymakers and organized interest groups; only two participants came from patient groups. Thus, the EGE certainly is an institutional innovation, since it raises the new topic of ethical questions. However, it still follows a traditional expert model because it involves only a small number of elite ethicists, scientists and jurists, who are also specialists on biotechnology issues. Public involvement is extremely limited in the EGE setting. The group seems to be aware of this problem and states in its Opinion No. 15 that "there is a need for continuing dialogue and education to promote the participation of citizens, including patients, in scientific governance, namely in the social choices created by new scientific development" (EGE 2000a: 20). However, it does not state how this objective can be achieved.

3.3.2.2 European Group on Life Science (EGLS)

In 2000 Commissioner Busquin installed the Life Science High Level Group, later renamed European Group on Life Science (EGLS), to "provide high-level advice on life science and associated technologies".[3] The group was in office until 2004 and was a typical elite scientific expert body. Besides the task of informing the Research Commissioner, it also aimed at supporting "science communication strategies, such as engaging in informed and pluralistic stakeholder debates on life science perspectives" (European Commission 2002b: 3).

---

[3] http://ec.europa.eu/research/life-sciences/egls/index_en.html (download 2007-12-10).

Under this mandate it organized the conference "Stem cells, therapies for the future?"[4] in December 2001 and invited 750 persons from 36 countries "including expert scientists, clinicians, politicians, industrialists, representatives of interest groups, patient support groups and religions, and interested private individuals" (ibid.: 6). According to Commissioner Busquin, this conference was "intended as an exercise in the governance of science" (ibid.).

> Illustrating its will to innovate and to maximize the impact of the debate, the discussion platform was web cast, and members of the public were invited to express their opinions and views via e-mail before, during, or after the forum. After each talk, there was a question-and-answer session, and the programme included two round-table discussions and one public debate. The aim was to air and publicize the current state of scientific knowledge and medical progress as a basis for further discussion (ibid. 6).

However, the results of this conference remain very vague. An official brochure states that it was "not possible to achieve total consensus, but widespread explicit or implicit agreement on several points" (European Commission 2002b: 33). This consensus, however, is formulated rather imprecisely: It was "generally accepted that the potential value of regenerative medicine involving stem cells is enormous". Moreover, there was "broad agreement that living human tissues, and human embryos in particular, should be respected". "Most participants felt that the use of (...) adult and umbilical cord blood stem cells was acceptable". "Many of those present seemed to feel that reproductive cloning was unethical and should be prohibited".

The rest of the conclusions are even more pleading and fuzzy: "[S]cience should not proceed in a vacuum", the "ethical debate requires input from many diverse elements of society, including the public at large and interest groups, not just scientists and technocrats", "there is substantial diversity in Europe (...) as concerns ethics in this field, and this may make it difficult to lay down common legislation", "public debate is required" (ibid.).

In contrast to these imprecise formulations, the EGLS's own recommendations presented at the end of the brochure are much more concrete and policy-oriented than the conclusions it draws from the Public Forum.

> "[R]esearch using cells from both sources should be actively developed and supported" (ibid.: 34). "None of the scientists were in favor of prohibiting stem cell research in general, or any particular type of stem cell research". (ibid.) "New lines need to be derived if this approach is to realize its clinical potential" (ibid.). "The EU should continue to support research with all sources of human stem cells, including human embryonic stem cells, to provide new clinical opportunities for therapy" (ibid.: 35). "Reproductive cloning (...) should be prohibited" (ibid.). "Derivation of human embryonic stem cells from nuclear transplantation (so-called therapeutic cloning) has not been achieved and appears to raise considerable difficulties, scientific as well as ethical" (ibid.). The group "agrees on the use of spare human embryos for the derivation of embryonic stem cell lines. (...) the Group insists that in those countries where research on human embryonic stem cells is allowed, it should be carefully regulated, peer reviewed, scientifically sound, directed towards substantial goals and ethically controlled" (ibid.).

[4] http://ec.europa.eu/research/quality-of-life/stemcells.html (download 2007-12-10).

Thus, although the conference was a major effort to involve the European public, though in a rather staged and unidirectional setting, on HESC research, it still was expert-oriented, just informative and vague in its outcome.

3.3.3 Informing the Public Directly and Indirectly

Another approach to involve the public is simply information provided on the Internet. The web page of DG Research provides information on policy papers and conferences taking place in Brussels on HESC research.[5]

The Commission also delegates this information task to all researchers. It "support(s) measures to help researchers become communicators and debaters, caring for the conditions in which all parties in society can be involved in embarking on new ways of collective learning" (Commission of the European Communities 2003a: 10).

3.4 Summary

In summary, the debate on HESC research continues to be mainly an elite concern, which does not include the general and concerned public, who feel that their national governments and/or the EU neglect their interests and concerns. However, and this is a major concern for policymakers, the issue of HESC research is potentially highly explosive, similarly to the topics of abortion, GMO, or BSE. Red biotechnology might be "at the beginning of (the) politicization curve" (Salter and Jones 2002: 334) which the regulation of GMO had followed in the 1990s. It is uncertain whether and to what extent the politics of HESCs will move in the same direction.

> Although the body politic of human genetics and health may at present appear to be unaffected by the political virus which has so virulently attacked green biotechnology, it would be unwise to assume immunity (Salter and Jones 2002: 337).

In contrast to the GMO debate, the controversy on HESC research primarily involves elite actors from EU institutions and traditional stakeholder organizations, which differ in their evaluations of values such as freedom of research, freedom of doing business, or the moral status of the human embryo. This group of elite actors includes officials of EU institutions and EU member states, top politicians, scientific and bio-ethics experts, as well as representatives of established, well-organized pressure groups and science, industry or church lobbies.

By and large, the debate is still characterized and decided by standard inter-institutional conflicts and dynamics within and among EU institutions. In particular, it involves a permissive Commission; the EU Parliament, which is split, but

[5] http://ec.europa.eu/research/fp6/index_en.cfm?p=1_stem_dialogue (download 2007-12-10).

ultimately permissive; a split Council of Ministers; and established pressure groups lobbying either for a permissive or restrictive cause.

The case of HESC research constituted a disruption in the process of setting up FP 6, which officials and policymakers in the institutions concerned tried to remedy by their customary administrative practices, such as comparing national regulations, compiling reports and gathering information. Commission officials and policymakers also tried to increase their expertise and legitimacy by involving elite researchers on life science, biotechnology and, particularly, stem cell research. Policy makers addressed the new and unruly topic of ethics by efforts to build up, integrate and harmonize expert bodies and expertise in bio-ethics. Thus, the Commission followed its well-established routines of expertise politics (Abels 2002: 3), i.e. scientists advising policy-makers in an ambiguous arrangement, in which the scientists assume different roles, acting as supposedly impartial experts and, simultaneously, as privileged stakeholders, thereby getting advantaged access to decision-making.

Moreover policy makers changed administrative routines by establishing an ethics review for funding decisions. In addition, Commission officials organized conferences, workshops, seminars and meetings, in order to get together with other policymakers and experts and establish common ground. Guiding principles in this process were formal powers and rules of procedure, but also informal rules and bargaining between the three major EU institutions concerned, i.e. within and between the Commission, the Council and Parliament. All these efforts mentioned so far were aimed at policymakers, politicians and experts.

However, what was the role of the public in all this? Since HESC is a very diffuse policy issue – in contrast, for example, to abortion – policymakers have been faced with the problem of "finding" a civil society with which they could negotiate, of identifying a public that would be able to participate. Participation proves to be difficult since civil society groups are less well organized than industry. For example, the European Parliament's Temporary Committee on Human Genetics had difficulty identifying relevant organizations (Salter and Jones 2002: 334).

In the decision-making process on HESC research within FP 6, participation and dialogue was mainly used as a means of informing the public. Participatory practices covered only a small share of activities and were never used in a consultative way or in actual decision-making. When citizens were addressed at all, they were invited to few, well-staged events, such as conferences or round tables, and they mostly remained in their role of an audience that was to be enlightened. Thus, despite the participatory language in some EU documents, the participatory opportunities regarding decisions on HESC research in the context of FP 6 remained extremely limited. If participatory practices occurred at all, they were few, organized from the top, expert-oriented, at best consultative and non-binding for decision-making.

Acknowledgements This paper is based on research for the project "Participatory Governance and Institutional Innovation" (PAGANINI), funded by the European Commission within the Sixth Framework Programme (Contract No. 505791). I would like to thank the Commission for funding and my PAGANINI colleagues for our fruitful discussions throughout the project. In particular, I want to thank Herbert Gottweis for granting me access to interviews he conducted with two civil servants, officials of the European Commission and of the Austrian government, on European HESC policy; I am also grateful to Peter Biegelbauer for his encouraging and practical comments and Silvia Plaza for her help in language editing.

References

Abels, Gabriele. 2002. Experts, citizen, and Eurocrats – towards a policy shift in the governance of biopolitics in the EU. http://eiop.or.at/eiop/texte/2002-019.htm (download 2006-11-26).

Adenberger, Caroline. 2005. Embryonic stem cell research policies: Focus on Europe. In Bridges – The OST's Publication on S&T Policy. Office of Science and Technology. www.ostina.org/html/bridges/article.htm?article=1247 (download 2005-09-29).

Capps, Benjamin. 2005. The human embryo, stem cell research, and the European Union. In Wolfgang Bender, Christine Hauskeller, and Alexandra Manzei (Hrsg.): Grenzüberschreitungen/Crossing Border. Kulturelle, religiöse und politische Differenzen im Kontext der Stammzellenforschung weltweit. [Cultural, religious, and political differences concerning stem cell research. A global approach.], 435–468. Münster: Agenda.

Commission of the European Communities. 2001a. European Governance. A White Paper. COM(2001)428.

Commission of the European Communities. 2001b. Towards a strategic vision of life sciences and biotechnology: A consultation document. COM(2001)454 final, Brussels, 4.9.2001.

Commission of the European Communities. 2003a. Life science and biotechnology – A strategy for Europe. Progress report and future orientations. Report from the Commission to the European Parliament, the Council and the European Economic and Social Committee. COM(2003)96 final, Brussels, 5.3.2003.

Commission of the European Communities. 2003b. Commission staff working paper. Report on Human Embryonic Stem Cell Research. SEC(2003)441.

Commission of the European Communities. 2004. Life science and biotechnology – A strategy for Europe. Second progress report and future orientations. Report from the Commission to the European Parliament, the Council and the European Economic and Social Committee. COM(2004)250 final, Brussels, 7.4.2004.

Commission of the European Communities. 2005. Life science and biotechnology – A strategy for Europe. Third progress report and future orientations. Report from the Commission to the European Parliament, the Council, the Committee of the Regions and the European Economic and Social Committee. COM(2005)286 final, Brussels, 29.6.2005.

EGE. 1998. Opinion of the European Group on Ethics in Science and New Technologies. Opinion No. 12. 23.11.1998.

EGE. 2000a. Opinion of the European Group on Ethics in Science and New Technologies to the European Commission. Opinion No. 15. Ethical aspects of human stem cell research and use. 14.11.2000.

EGE. 2000b. The European Group on Ethics in Science and New Technologies to the European Commission. Adoption of an opinion on ethical aspects of human stem cell research and use. Revised edition, January. Paris, 14.11.2000.

EGE. 2001. The European Group on Ethics in Science and New Technologies to the European Commission. General report. On the activities of the European Group on Ethics in science and new technologies to the European Commission 1998–2000.

European Commission. 2001a. Stem Cell Research at European level. Meeting report 13–14 September 2001, Brussels. Research Directorate-General, Directorates E and F for Life Sciences. Prepared by G. Joliff-Botrel and L. Matthiesen, November.

European Commission. 2001b. Survey on opinion from National Ethics Committees or similar bodies, public debates and national legislation in relation to human stem cells. Quality of Life Programme. May, Brussels.

European Commission. 2002a. Life sciences and biotechnology. A strategy for Europe. Communication from the Commission to the European Parliament, the Council, the Economic and Social Committee and the Committee of the Regions. COM(2002)27, Brussels.
European Commission. 2002b. Stem cells. Therapies for the future? Luxemburg.
European Commission. 2004. Survey on opinions from National Ethics Committees or similar bodies, public debates and national legislation in relation to human embryonic stem cell research and use. In EU Member States. Directorate E, Biotechnology, Agriculture and Food, ed. Line Matthiessen-Guyader, Vol. I. July.
European Council. 2000. Lisbon European Council 23 and 24 March 2000, Presidency Conclusions. www.europarl.europa.eu/summits/lis1_en.htm (download 2007-12-10).
European Parliament. 2003. European Parliament legislative resolution on the proposal for a Council decision amending Decision 2002/834/EC on the specific programme for research, technological development and demonstration: "Integrating and strengthening the European research area" (2002–2006) (COM(2003) 390 – C5-0349/2003 – 2003/0151(CNS)). P5_TA(2003)0506. http://www.europarl.europa.eu/sides/getDoc.do?pubRef=-//EP//NONSGML+TA+P5-TA-2003-0506+0+DOC+PDF+V0//EN (download: 2011-06.23).
Forum National Ethics Council. http://ec.europa.eu/research/science-society/page_en.cfm?id=3161 (download 2007-12-10).
Gottweis, Herbert. 2003. Embryos for Europe: Emerging strategies and institutional situations. Paper prepared for Workshop 21 "Assessing Emergent Forms of Governance: European Public Policies Beyond the Institutional Void". ECPR Joint Session of Workshops, Edinburgh, 28th March–2nd April 2003.
IP/00/1293. 2000. The European Group on Ethics makes public in Paris its opinion on "Ethical Aspects of Human Stem Cell Research and Use". Paris, 14 November 2000.
IP/00/15019. 2000. Philipe Busquin reacts to the UK Parliament's vote on therapeutic cloning. Brussels, 20 December 2000.
Interview. 2002. With Civil servant, European Commission. Interviewed by Herbert Gottweis, 13 January.
Pichler, Franz. 2005. Science and Human Dignity. Ethical questions on the funding of human embryonic stem cell research in the Sixth Framework Programme of the European Commission for research, technological development and demonstration (2002–2006). In: Innovation, Vol. 18, No. 2, 261–271.
Romeo-Casabona, Carlos M. 2002. Embryonic stem cell and therapy: The need for a common European legal framework. In: Bioethics, No. 16, 6 November, 493–503.
Salter, Brian. 2005. The global politics of human embryonic stem cell science. Global Biopolitics Research Group, University of East Anglia, Working Paper No. 2, April. www.globalbiopolitics.org.
Salter, Brian, and Jones, Mavis. 2002. Regulating human genetics: The changing politics of biotechnology governance in the European Union. *Health, Risk and Society* 4(3):325–340.

Chapter 4
The Politics of Human Embryo Research in Poland

Teresa Kulawik

4.1 Introduction

Epoch-making developments in biotechnology and biomedicine have spawned a rapidly growing field of research on biopolicy regulations in different countries (Abels et al. 2003, Bleiklie et al. 2004, Gottweis 1998, Jasanoff 2005, Russel and Vogler 2000). As yet, analyses of postsocialist democracies have remained fairly scarce, as the bulk of these studies focus on Western countries. It is understood that postsocialist countries clearly do not form one uniform block and the ways in which these countries address the challenges of new technologies are often quite different (Just 2008, Sandor 2003). According to a survey by the European Commission, Estonia, Hungary, Latvia and Slovenia have laws that permit research activities on embryos; whereas Lithuania, Poland and the Slovak Republic prohibit the procurement of human embryonic stem cells (ESC).[1] As we shall see, adding Poland to the latter is incorrect. Such a mistaken interpretation could partially be attributed to Poland's fairly complex legal situation involving the so-called "conceived child" (*dziecko poczęte*), which has been the legal term for the embryo since the 1990s. Although the "conceived child" is endowed with a high moral as well as legal status

---

This chapter is a revised version of a paper, which was presented at the European Consorium for Political Research Conference in Budapest September 2005 as well as at the Fifteenth International Conference of the Council for European Studies in Chicago/USA March 2006. I would like to thank the discussants and participants for their valuable comments. This study is part of a research project on gene technology, democracy and deliberation in Germany, Poland and Sweden, which was supported by the Swedish Research Council. The author would like to thank all her interview partners in Poland. She is especially indebted to Eleonora Zielińska, who has considerably eased the task of collecting material, for her support. She would also like to thank Monika Zima and Maria Cichocka for their research assistance. Last but not least she would like to thank Erika Doucette and the editors who have turned the manuscript into a printable English text.

[1] http://www.europa.eu.int/comm/research/biosociety/pdf/mb_states_230804.pdf

T. Kulawik (✉)
Södertörn University, Stockholm, Sweden
e-mail: teresa.kulawik@sh.se

P.T. Robbins, F. Huzair (eds.), *Exploring Central and Eastern Europe's Biotechnology Landscape*, The International Library of Ethics, Law and Technology 9,
DOI 10.1007/978-90-481-9784-2_4,

in the penal and civil law codes, regulations concerning reproductive and gene technologies are still lacking. It is rather controversial among politicians and experts whether the existing medical law and the legal framework concerning abortion, which includes a ban on "experiments" on the "conceived child," can be applied to these new technologies.[2] In conjunction with Poland's entrance into the European Union its government was requested to make a statement clarifying their stance on financing research on ESCs within the EU's 6th framework program. The left-wing government had agreed to finance research under certain conditions.[3] This statement implies that research on human ESCs conforms to Poland's regulations in effect.

Political controversies surrounding new technologies are regarded as both a challenge and a chance for renewing representative democracies. This pertains to the concept of deliberative democracy as well as to different technology assessment procedures involving the public (Elam and Bertilsson 2003). Sheila Jasanoff (2005: 6) asserts that without taking into consideration the politics of science and technology "democratic theory cannot be articulated in satisfactory terms today."

This leads us to ask to what extent such claims are relevant for postsocialist democracies. Despite the existence of environmental movements in postsocialist states, conflicts concerning development and application of technologies have not had high priority on the political agenda (Przestalski et al. 2001). In contrast to many western European countries, where in recent years value conflicts have been primarily triggered by the development of new technologies, in Poland, after 1989, the overarching ethical controversy centered over the "old" abortion question. Nevertheless, even if these new technologies appear to be less significant in the new democracies, they can indeed serve as a "window", to use Jasanoff's words (2005: 5), for looking into the architecture of postsocialist democratic governments. The aim of this study is to explore how the young Polish democracy copes with the challenges of the biomedical policy issues. In particular I will examine how the high moral and legal status of the embryo, that motivates the restrictive abortion legislation, is compatible with a politics of non-decisions concerning biomedical practices, which involve the procurement and handling of human egg cells. The lack of regulations implies that Poland de facto functions as a country with a permissive policy design in that policy field (for comparative classifications see Bleiklie et al. 2004).

The most frequent argument in order to account for Polish reproductive politics is the reference to the role of the Catholic Church (for examples see Just

[2]Andrzej Rzepliński is the name of the contact person mentioned in this survey. In 2005, in the Senate, the second chamber (or House) of the Polish parliament, Rzepliński's appointment as Commissioner for Civic Rights was overturned by the votes of the left party SLD. According to the Polish daily *Gazeta Wyborcza*, 23–24 June 2005, p. 3. SLD was opposed to Rzepliński, among other reasons, because of his restrictive stance on the protection of the "conceived child."

[3]According to a resolution of the Council of Ministers on 13 January 2004, this information follows a suggested recommendation that was presented at a meeting of the EU Council on 3 December. http://www.kbn.gov.pl/komorki_macierzyste/20040126.html (last accessed 20 March 2005)

2008). The Church could indeed be an important factor, however the role of the church does not explain the inconsistent policy pattern, nor does it suffice as a variable to predict policy variation between countries, as the cases of France, Spain and Hungary demonstrate, where Catholicism has not prevented the legalization of embryo research. That religion as "belief" might have much less predictable power about concrete policy regulations, as commonly assumed, is also illustrated by recent population polls on stem cell research in the European Union. Countries with the highest level of approval of ESC research are both Protestant and Catholic (Gaskell et al. 2006: 35–36). Assuming Catholicism as religion/world-view or institution were important, the means and mechanisms by and with which it asserts its influence in policy-making processes would still need to be investigated.

I argue that the puzzling pattern of Polish policy design can be best analyzed when applying a discursive institutionalist approach which, like no other school in policy studies, theorizes and investigates the complex dynamics of institutional arrangements, actor constellations, and political discourses from a long-term perspective (Immergut 1998, Schmidt 2008, Kulawik 2009a). Discursive institutionalism helps to question culturalist approaches, which too easily explain certain phenomena in the consolidating democracies, in terms of "mental legacies" of the communist era or, in the Polish case, with Catholism. No doubt religion and cultural legacies play a role, but their influence on politics has to be accounted for and not taken for granted (Brier 2009). In addition, the impact of institutional arrangements on citizens' behaviour should not be underestimated (see Holmes 1996). In contrast to rational-choice institutionalism, the framework of discursive institutionalism helps to understand why politics in democratic Poland evolve around cultural cleavages rather than socio-economic interests.

In short, I argue that the abortion struggle has definitively shaped the public sphere and established a hegemonic paradigm of "public morals", thus hindering policy-oriented debates on bioethical dilemmas. This is compounded with a limited policy-making capacity, which makes it extremely difficult to decide on contentious issues and reinforces strategies of evading the issue altogether. The Polish decision-making system is hardly prepared for taking on these debates. An attempt to activate an "informed" discussion was made by the leftist government in 2004 enacting a so-called "societal consultation" on the use of human embryonic stem-cells (ESC) for research, which was triggered by Poland's accession to the EU. Although this consultation process was at least partly geared toward attaining a mutual understanding, it was not able to find a compromise between the advocates and adversaries of human stem cell research, which would make legislature more feasible.

The article starts with a presentation of the puzzling features of Poland's legal regulations in the field. It continues with an analysis of the structuration of the public sphere through the discursive struggle on abortion and its consequences for addressing biomedical practices, which involve the procurement and handling of human egg cells. The third section examines Poland's policy-making capacity. The fourth section traces the societal consultation process concerning embryonic ESC research carried out under the label "Life for Life". The final section discusses the study's findings.

4.2 Obscured Policy Regulations

As I have already noted, Poland has yet to codify a law that explicitly covers assisted reproductive technologies including the procurement and handling of human gametes. Regulations pertaining to the legal status of the embryo are laid out mainly in two areas of Polish law: in abortion law and the Act on the Medical Profession. The complexity of the present legal situation is a result of the lack of clarity of the legal term "conceived child" (*dziecko poczęte*) that was introduced in 1993, and of the innumerable revisions made in the 1990s.

As early as 1956, following the Soviet Union, Poland had a permissive abortion law that allowed for the termination of pregnancy based on medical, legal and social grounds.[4] In 1993, the Polish Parliament passed the Law on Family Planning, Legal Protection of Human Fetus and the Conditions of Admissibility of Interruption of Pregnancy (hereafter called the Act on Family Planning).[5] In addition to severely limiting the availability of abortion by repealing the social clause as grounds for abortion and prohibiting abortions in private clinics, the Act radically changed the legal status of the embryo. Previous legal provisions only indirectly protected the embryo under the regulations on the protection of the health of pregnant women. The Act on Family Planning devised the "conceived child" as an independent legal subject. The preamble to this law states that "the right to life and health are fundamental rights to be secured by the state". This includes the mention in Article 1 that "the right to life shall be protected including the prenatal stage thereof, to the extent laid down in the law". Article 6 of the Act is a regulation added to the Civil Code, which provides that "a conceived child shall likewise enjoy legal capacity". In addition, "causing the death of a conceived child", as is mentioned in Article 7–2, introduces a new criminal offense that conceives of abortion in analogy to homicide, that is, as a principally penalized act of killing that is exempt from punishment under certain conditions. This law determines that the "mother of a conceived child" is exempt from prosecution if she terminates her pregnancy illegally; furthermore, an abortion performed by a doctor is only legal if medical and criminological indications are at hand.[6] Article 7–1 of the Act on Family Planning adds regulations to the Penal Code on permissible risks involved in treatment and "experiments" and determines that the "conceived child" may not be subject to any activities other than those intended to protect the life and health of the child or that of the mother". Article 7–4 introduced a hitherto unknown type of criminal offense, namely "causing bodily injury to a conceived child or the impairment of the health of a conceived

---

[4] O warunkach dopuszczalności przerywania ciąży. Dz.U. (1956) no. 12, item 61 (Dz.U. – *Dziennik Ustaw* – Official Journal of Laws, hereafter Dz.U.)

[5] O planowaniu rodziny, ochronie płodu ludzkiegu i warunkach dopuszczalności przerywania ciąży, Dz.U. (1993) no.17, item 78.

[6] Art. 149a and 149b of the Penal Code (PC). The medical clause covers the case that the life of the mother is endangered or the child has a genetically inherited disease. The punishment is up to 2 years in prison.

child that endangers its life". The point that follows deems that if such an act were to result in the "death of a human being" that it is punishable by 1–10 years in prison.

As its name already indicates, this law was codified as the result of a compromise between two quite contrary bills. One of the bills was proposed by the clerical-nationalist as well as center-right-wing post-Solidarnosc parties and was presented as "legal protection for the conceived child". This bill contained the new criminal offense mentioned above and set out to allow abortion only in cases where the health and life of the pregnant woman are at risk.[7]

Postcommunist leftists presented a counterproposal under the heading "On Family Planning, Legal Protection of Human Fetus and the Conditions of Admissibility of Interruption of Pregnancy". This became the name of the law that was passed in 1993, which was however much more restrictive. The bill's objective was to engender responsible planned parenthood.[8] In addition to upholding the existing system of clauses its provisions included providing family planning centers and contraceptives free of charge. The interesting thing about this bill is that it included a regulation on assisted reproduction that addressed the right to infertility treatment. According to Article 13 spare "human embryos" resulting from this treatment may neither be regarded as property nor as an object of trade. Yet how spare embryos were to be dealt with was to be covered by a different law. In addition to prohibiting commercialization of the human embryo, the proposal also prohibited cloning and the production of chimera (Article 23–24). The passed Act on Family Planning included regulations on sex education and assistance during pregnancy; however the issue of assisted reproduction still remained open. Assisted reproduction was neither included in this bill nor in any subsequent amendment to the law.

The Act on Family Planning of 1993 aimed to codify a comprehensive protection of life and health of the "conceived child" in Poland; however its reach of coverage remained unclear. Shortly after its enactment, legal commentaries questioned whether it covered the entity in-vitro or only in-vivo. Eleonora Zielińska supported the standpoint that protection – particularly within the Penal Code –was only valid in-utero, which is an interpretation the majority of the commentators did not agree with (Zielińska 1995, 2005, Jarosław and Wróbel 1993). Since 1993 the Act of Family Planning has been repeatedly amended. In 1996, the postsocialist leftist government replaced the concept of protecting the "child from the moment of conception" with a more generally formulated regulation that provided protection in the "prenatal phase".[9] However, the Constitutional Tribunal objected, deeming the bill constitutionally incompatible. The Tribunal found the regulations involving the social clause to be too imprecise, and thus incompatible with the right to life

[7] Sejm of the Republic Poland (1st Chamber of the Polish Parliament, hereafter called SejmRP), print no. 190 (25 March 1992).

[8] SejmRP print no. 195 (30 March 1992).

[9] O zmianie ustawy o planowaniu rodziny, ochronie płodu ludzkiegu i warunkach dopuszczalności przerywania ciąży, Dz.U. 1996, no.139, item 646.

laid out in the constitution. This decision was controversial, because the so-called "small constitution" in place at the time did not explicitly contain any mention of the right to life (Zielińska 2000). The Tribunal's verdict was that the constitutional protection of the lives of all humans including the embryo should be derived from the rule of a democratic state of law.[10]

In 1999, following the post-Solidarność parties' successful election and the change in government, these legal regulations were again revised. The language reverted back to using terminology, such as the "conceived child" and to concepts of criminal offences of "causing death" or "causing injury and endangering the life of the conceived child".[11] At the same time, the Act on the Medical Profession was amended. This 1997 law imposed comprehensive regulations on research on human beings. The Polish vocabulary used to describe this research is literally "experimenting on human beings", whereas a distinction between therapeutic and scientific experiments is made. Although the law passed in 1997 had deemed research on pregnant women and minors to be subject to very strict risk assessment, there is no mention of the prenatal phase.[12] The 1999 amendment added the category "conceived child" to point 3 of Article 26, which was a general prohibition to take part in scientific or medical experiments for "soldiers, prisoners and those without the power of decision under law".

It would seem that this legislation established in Poland provides a comprehensive legal protection of life in the prenatal phases and prohibits research on embryos ex-utero. This is Elenora Zielińska's (2005) conclusion. However, the situation is not as clear as it may seem. The problem is that although legislature seeks to provide general protection for the "conceived child", this term is never explicitly defined. As such, this becomes an indefinite legal term, the inherent meaning of which depends on the recognition of certain – disputable – determined factors. They may be "intended" by the legislature, but they are not laid down in the law. One such factor in this case is that "conception" is defined as a moment and not a process; that the "moment" is the fusion of the egg cell and the sperm, which excludes, for example, interpretations based on the nidation. Marek Safjan, former president of the Constitutional Tribunal and medical law expert who paved the way for adopting the concept of protecting the "conceived child" in Poland, contends that infertility treatment, as a "a common practice" in Poland, still remains "outside Polish law". Ultimately, this law only indirectly implies prohibitions such as that of creating embryos for research purposes (Safjan 2003). In an interview he stressed that "the Penal Code limits its protection to that of the child inside the body of the woman" (Safjan 2000, 2001).

---

[10]The Ruling of the Constitutional Tribunal of the 28 May, 1997 K26/96, here: http://www.pdi.net/~polfedwo/english/constrib.htm.

[11]O zmianie ustawy – Kodeks Karny oraz ustawy o zawodzie lekarza, Dz.U. (1999) no. 64 item 729.

[12]O zawodzie lekarza, Dz.U. (1997) no. 28 item 152.

Because of its effort to provide comprehensive protection of the right to life, the concept of the "conceived child" consciously neglects differentiation between in-vivo and in-vitro. This allows the government room for interpretation regarding the legal situation. This may elucidate how it was possible for the Polish Government to speak in favor of a moratorium on reproductive and therapeutic cloning in the United Nations in 2003, and then in 2005, vote in favor of differentiating between these two forms of cloning and accept the use of ESCs under specific conditions (UN 2005).

> Poland is of the opinion that any use of human embryonic stem cells, including for the purposes I have mentioned, should be permitted only if the following conditions are met: stem cells and stem cell lines are obtained from reliable and documented sources; human embryos used to obtain human stem cells or to create stem cells lines are supernumerary cells, meaning embryos which were created in the process of in vitro fertilization aimed at initiating a pregnancy but are no longer aimed at achieving the said goal; the donors of embryos have expressed in a written form their free and unequivocal will for their embryos to be used in a particular way; anonymous donors of embryos are excluded and the personal data of donors, including their genetic data, is subject to full protection; the donors of embryos were not given or promised any pecuniary or material benefit. If any of the aforementioned conditions are not met, Poland is opposed to any kind of use of embryonic stem cells.

4.3 The Public Sphere and "Public Morals"

The conflict on abortion, also known as the "abortion war," was one of the foundational struggles of Poland's democratic government. The debates were not only on abortion, they also included fundamental issues surrounding democratic citizenship, the boundaries between private and public, gender differences, the linkage between morals, ethics and law, and last but not least, approaches to dealing with the country's communist past (Zielińska 2000). It should be mentioned that abortion was on political agendas during the transition phases of most postsocialist states. While this freed them from "communist" prohibitions in some countries, it prompted more restrictive regulations in others (Gal and Kligman 2000). No matter which regulations were passed, reckoning with the former "regime of injustice" and the installation of a new "proper" order of the body politic was carried out on women's bodies and a redefinition of life itself. This confirms the significant role that politics of life play in re-imagining national communities, as this is constitutional to nation-building processes as well as representational of critical junctures in societal change. Therefore I do not consider abortion as a "substitute" issue where "wider concerns" (Kramer 2009: 82) are somewhat accidentally debated. Instead bodily and reproductive issues have throughout history been a major discursive terrain for the articulation of the relation between individual subjects, collectivities and social order (Planert 2000, Youval-Davis 1998). Gendered and racialized bodies have served as key markers of belonging and membership in a national community. Assumptions about bodies are among the most important tools in distinguishing full from lesser citizens (Bacchi and Beasley 2002: 325). "Big" debates about bodily issues are concerned with the very constitution of the body politic. Therefore it was not accidential that throughout the 1990s the debates surrounding the drafting of Poland's constitution and of the abortion law ran parall.

As early as in the 1980s, there had been attempts to reform the permissive abortion laws in Poland. Although these attempts were spearheaded by the Church these were widely supported by the Solidarność movement. The first revision of the bill was introduced in Parliament in 1989 when Poland was still under communist rule. Massive protests during the decisive period of elections in June 1989 put it on hold. The struggle on abortion law continued throughout the 1990s, mobilizing Polish society and occupying the political agenda like no other issue. There were extensive pro-life and pro-choice demonstrations, NGO actions, events and discussions and several political initiatives across all parties were formed (Zielińska 2000).

A decisive moment in this process was the invention of a new language. Until then, abortion had been framed in commonly used neutral medical terms. These were replaced by profoundly value-laden ways of speaking about abortion. For example, the embryo (*płód*) became "conceived child", "little boy/girl" or "child in the mother's belly" and the pregnant woman was referred to as a mother. The operation was no longer done by a doctor, but by an "aborter". It no longer took place in a hospital, but in an "abortion chamber", which aimed to evoke an analogy to a gas chamber (Matuchniak-Krasuska 1991). This change in language established rules of political discourse. The political opponent was not an adversary, but was instead considered an enemy or even a criminal. Insinuating that your opponent is in favor of a legal form of murder not only severs the lines of communication, it also leaves no room for negotiation.

It is remarkable how similar these politics of language are to the semantic strategies of the so-called *nowomowa* (new speak). This "new speak" had been employed by the communist powers and was then astutely exposed by Polish intellectuals as part of the "culture of the lie" (Kaluza 1998). Ironically the discursive parameters of the "culture of truth" of the Polish opposition not only mirrored some of the dictatorial semantics – such as the divide between "them" and "us" – but also prepared ground for "moral aggression" as the dominating pattern of Poland's political discourse in the 1990s (Brier 2009). The argumentative framework of the Polish opposition movement was foremost a moral one. Solidarity's ethos centered around moral categories which since the 1970s received major inspiration from Catholicism: the dignity of the human person and human rights derived thereof as well as the vision of the Polish nation as a "moral community". The catholic interpretation of human rights, developed in the aftermath of the Second Vatican Council, was attractive due to its transcendent character: dignity was perceived as a given, as part of a "natural order"; hence undisputable. In addition Catholicism also provided a powerful master narrative of the Polish nation state in which the nation's survival was presented as depending on its religious fidelity.

The invigoration of a democratic public sphere in Poland after 1989 was deeply marked by the legacy of the political cleavages of the 1980s: *First* the translation of political conflicts into moral ones; *Second*: the unresolved tension between Solidarity's two visions of nationhood and citizenship. Solidarity articulated an ideology that could be interpreted in two ways: (1) as a civic project based on inclusive universalism and a pluralist representative democracy legitimized foremost by its institutional arrangements and (2) as an ethnic project based on a cultural, homogenous conception of Polishness and a democracy founded on the idea of an authentic common will. Within both paradigms the Round Table could be conceived accordingly: either as passage towards a peaceful, negotiated transition or as failure of the "re-appropriation" of Polish politics through the truthful "moral community".

The struggle over abortion was heavily imprinted by the discursive frameworks and its conflicting visions inherited from the Solidarity period. The characteristic style of the discussion was, in line with the polarized "we versus them" schema, one of exclusiveness and unconditionality. Representatives of both sides constructed their position as the only righteous, true and feasible one. Those in favor of prohibition linked abortion to the crimes committed under communism and deemed the protection of life the most important foundation for a righteous and true democracy. They argued that this question was connected to realizing the basic principles Poland had fought and died for, namely human dignity, freedom of religion and honesty. The pro-life advocates argued with the Pope's scriptures as well as with scientific evidence. Their argument was not founded on the doctrine of animation; instead they turned to embryology. There is a simple answer to the question of the origin of life, according to a Member of Parliament. It was said that by now it should be common knowledge that when the sperm and egg unite a complete set of genetic information is created, which then only needs to unfold in the phases that follow. Thus, for him, "from the perspective of genetics, there is nothing to be debated" (Pawlik 1991: 133). This quote reflects a typical structure of the right-wing discourse, which is also present in the constitutional debate. It conceives democratic politics in Poland as a *recreation* of a natural order based on certain "facts" which are withdrawn from discussion (Brier 2009: 67). It is remarkable, and will be demonstrated in more detail below, that within the Christian right-wing framework values are increasingly derived from scientific "facts", rather than motivated by ethical reasoning itself.

The abortion debate is embedded in the historical narrative of Poland's heroic epic of the people's struggle that was avidly supported by the Catholic Church. The abortion debate not only addresses efforts to come to terms with the communist past, moreover it seeks to project a new national identity, which – in line with the heroic narrative – aspires to position Poland as spearheading a "civilization of life" as opposed to a "culture of death". As one senator put it: "Our country may not serve as an economic role model, in relation to the technologically advanced Europe we can provide a model of moral order" (Fuchs 2003: 183). Gender differences are integral to that moral order in which women are situated as lesser citizens, primarily as reproducers without rights to bodily integrity. The change in language referring to "pregnant women" as "mothers" is symptomatic here. Pro-life discourse goes hand in hand with a revitalization of motherhood as metaphor and stereotype in the public space. Historically the heroic Polish Mother has been central to Poland's national identity, representing national culture and the nation's suffering and survival. In the abortion debate after 1989 selfless motherhood is reinstalled as symbolic figure within the nationalistic framework and as part of the "re-appropriation" of Polish politics.

According to Agnieszka Graff (2009) the object of interest are not "unborn children" but the Polish nation and the policing of its boundaries. Those who talk of "rights" are perceived as anti-Polish intruders. Within that nationalist anti-choice discourse women are simultaneously reduced to bodies and also disembodied at the same time. Actual women figure as "wombs" and women's lived bodily experience is disregarded. The protection of "life" is conceived as the highest value of the democratic Polish state, while women's bodily integrity is dismantled.

The discussion on abortion shows that a central presupposition of a debate on ethical dilemmas and complex issues surrounding biotechnology regulations cannot be taken for granted. Value pluralism is not generally acknowledged in the Polish public sphere. This is closely related to another core principle of modern secular democracies, namely to the distinction of morals from jurisprudence. In Poland this principle that renders illegitimate any direct conferral of value judgment into universally binding laws is highly disputed. Clerical-nationalists and conservative intellectuals branded this separation by calling it "relativism" or "liberal totalitarianism" (Górski 2002, Środa 2005).

The abortion struggle has decisively shaped the parameters of political discourse in Poland, where politically controversial issues are readily fought out in a polarized and personalized manner. Discrediting political opponents, little consideration of opponent's arguments and compromises are the main characteristics of political debates in Poland (Fuchs 2003: 90–92). At the same time the abortion debate has staked out significant discursive claims in the realm of the politics of life. Reframing has been successfully employed as a strategy. The media – with the exception of the feminist press – now commonly uses the term "child" instead of embryo. In relation to questions of ethics and the politics of life, the Catholic Church has become an established power that politics are not able simply to circumvent. After the transition period, during a time when attempts to assert political influence directly were not tolerated by the bulk of Polish society, the Church changed gear and switched to more subdued policy-making strategies. The final communiqué regarding Poland's accession to the European Union provides an example of how well this functions under a leftist government. The communiqué, the so-called "Declaration of the Government of the Republic of Poland on Public Morals", assured that EU law would not have the power to interfere with the Polish regulations of "moral importance or concerning the protection of human life" (Hierlemann 2005: 211).

This did not stop the Polish government from taking certain stances, for example, as they had done on the question of research on human ES cells in the United Nations resolution and within the 6th framework program of the European Union. In their domestic policy, the Polish government tried to downplay these statements and relativize them in their responses to queries in Parliament. Several interpellations recorded in the Parliament made use of *nowomowa* semantics. In a question on the statement on the 6th framework program of the EU, a Member of Parliament compared procuring stem cells from the "body of the unborn child" with

> Nazi eugenic practices that negate the natural law (characteristic of all totalitarianisms: Hitler's, communist and liberal) and negate the unconditional protection of life, specifically of such persons who are not able to protect themselves from strong forms of aggression.[13]

This statement from the Sejm offers insight on how difficult it might be to enact legislation on assisted reproduction and embryo research. This could also be seen as part of a new culture of lies, to which abortion already belongs.[14] Without regulations or control on practices pertaining to assisted reproduction so-called "spare embryos" could easily become an export commodity in these times where demand is high on the international market.

4.4 A Limited Policy-Making Capacity

From a historical institutionalist point of view Poland is not only an excellent case that proves "history matters", but it is also a well suited case for studying the intersection of institutional settings, political identities and strategies for action. Ironically Poland's constitution entails a separation of powers, the main purpose of which is to limit the government's power of decree rather than to ensure the capability to govern in the public interest. This lays the legal groundwork for a political decision making process that endows several different political organs with veto rights. The second chamber (Senate), the president and the Constitutional Tribunal in Poland can veto the political rulings made in the Sejm, the first chamber of the Polish Parliament (Raciborski and Wiatr 2005, Ziemer and Matthes 2004). This is not only troublesome, because it curtails the enforcement of political decisions, but it is also problematic because it consolidates mindsets that prefer obstructing political processes over making compromises. The party system's fragmentation and instability, which are largely due to Poland's electoral system, add to the problem. This makes it rather difficult to form a stable government. In the 1990s there was an increase of polarization in the Polish party system, which revitalized the socialist legacy of a bipolar "we versus them" conception of the public space (Kubik 2003). This has had grave effects on processes involved in formulating political demands and objectives. Consequently, policy accountability tends to be low (Kulawik 2009b).

In the Polish case, the intense lines of conflict around cultural cleavages rather than on socio-economic interests do not make policy-making any easier. The decisive factor in determining if a political party is right or left-wing is not necessarily their economic agenda, but rather their position regarding Poland's past and the role

---

[13] SejmRP, interpellation, no. 8325 (4 October 2004).

[14] According to Hierlemann (2005): 212 there were 159 legal abortions and an estimated 200,000 illegal abortions in Poland in 2002.

of the Church. The extent to which the past influences present-day politics is illustrated by the fact that, despite many changes in government, Poland has never had a government coalition that includes parties from the post-Solidarity and the post-communist parties. However, this is not a common practice in other post-communist states either (Raciborski and Wiatr 2005: 225). The limited capacity for policy-making and the exceptional role that symbolic–religious questions play for political divisions elucidate the great challenge that finding regulations for contentious issues such as embryo research pose for Poland's political system.

Finally, I would like briefly to draw attention to the particularly important link between knowledge, experts and the political system within the relevant policy field. I gather from the sparse literature on the topic that the interactions between science, technology and politics are characterized by a low level of reflexivity. There is no long-standing practice of technology assessment or related interdisciplinary research. Like most modern states, Poland also has an advisory system which includes experts in the policy process but this is rather "sporadic" as Wieslaw Staskiewicz (2008: 33) states. According to Krzysztof Michalski (2003), consultancy practices on ministry level remain opaque and in most cases when a bill is drafted a ministry singles out experts who are consulted as "oracles". In the Sejm and Senate the chairs of the committees are responsible for consulting external experts in the legislation process.

Thus advice through experts in policy-making occurs, but foremost on an ad hoc basis because rules concerning the appointing procedure are lacking. The Polish political decision-making process does not provide the possibility of forming a parliamentary inquiry or expert commission to investigate complex policy issues, to propose legislation and to stimulate public debates. Since the early 1990s, there have been demands to establish a National Council on Bioethics, which would fulfill this function in the policy field of biomedicine and life sciences (Safjan 1992, Zielińska 1999). In the meantime several attempts have been made to install such a council, all to no avail.

4.5 Life for Life – Public Consultation on Human Stem Cell Research

The fact that Poland has not yet passed policy regulations on biomedicine does not mean that it is not a topic of public interest. After the birth of Dolly the sheep (at the latest), media and scholarly discourses took place to discuss the opportunities and risks of human biotechnology (Komitet Etyki 1997, Twardowski and Michalska 2000, Nauka 2003, Döring and Zinken 2005, Gazeta Wyborcza 2004). Since then a considerable number of books have appeared on the issue, including numerous translations of the works of Francis Fukuyama, Jürgen Habermas and Peter Singer. The bulk of the publications by Polish authors is situated within a Catholic context (Katolo 2000, Chyrowicz 1999, 2000, Bołoz and Höver 2002). These authors are largely members of Catholic think tanks, such as the Catholic

University, Lublin, and Cardinal Stefan Wyszyński University in Warsaw, which now houses a Center for Bioethics. Therefore, to assert that there is no public debate on the issue in Poland would be a false generalization. However, what is still lacking in broader public discussion is a policy-oriented examination of the ethical dilemmas. The prevailing discussion is led by natural scientists, doctors and philosophers and addresses the possible opportunites and "dangers" – the term "risk" is not commonly used – of genetics for "civilization" (Medycyna wieku rozwojowego 1999, Przyłuska-Fiszer 2005, 2001). It could be said that the terms of debate largely originate in the early phase of technology studies before they were seized by the constructivist turn. In view of the newness of the phenomenon at stake, voices that question the appropriateness of traditional categories remain scarce (Łuków 2001).

The Polish debate is characterized, first, by the dominance of deontological positions in the discussion on ethics, in which Christian ethics based on the writings of the Polish Pope has an obvious place (Jacorzyński and Kozłowski 2005: 474–475). A second trait is the use of a historical frame of reference. More often than not, historical references produce dubious analogies, such as "neo-Mengelism", named after the Auschwitz doctor, Josef Mengele. Respected scholars and writers, not only sensationalist reporting, frequently evoke such images in their publications. In this vein, a physician on TV described cloning as a way of breeding (hodowanie) humans as if organs were cut off in small pieces (MNI 2003: 3, quoted by Maciej Żylicz). Marek Safjan (2004) attributes the lack of a broader "authentic" discussion to the fact that "intellectuals" have neglected to provide the public with sufficient information. Between the lines of this critique of the opinion-makers it is clear that the absence of an informed discussion reflects the average citizen's general lack of knowledge. I would like to argue that the above analysis of the abortion debate leads rather to the presumption that at the core of the problem of an "authentic" discussion on bio-policies is not the average citizen, but the way in which Poland's political system and public sphere function.

An attempt to initiate an "informed" discussion was made by enacting the so-called "societal consultation" on the use of human ESC for research, which commenced at the time of Poland's accession to the EU in late 2003 and ended in summer 2004. At the onset, the Polish government had assigned the State Committee for Scientific Research (Komitet Badan Naukowych, hereafter KBN) to respond to the request of the EU's 6th Framework Program. The KBN generally favored research on stem cells, but sent a reminder that it was necessary to provide alternatives for research on embryonic cells and to conduct an "objective and constructive" discussion (KBN 2003).[15] The European Commission was not

[15]Originally the committee combined the role of a ministry of science and technology with that of a research funding agency. The work of the committee was headed by its chairman, the Minister of Science, and 12 representatives of the scientific community in Poland, elected through general election by all academics holding a doctoral degree. Information available at <http://kbn.icm.edu.pl/en/science/kbn.htm>, last accessed: 24 June 2004. Meanwhile the committee has been dissolved and the Ministry of Science reorganized.

satisfied with simply receiving a statement by KBN and, in line with its policy, the European Commission put in a request for a broader public dialogue. Following this, the cabinet designated by the Ministry of Science and Information Technology administered a resolution for a consultation with seven program points[16]:

(1) A conference with natural scientists, philosophers of ethics, legal experts, parliamentarians and journalists (December 2003).
(2) A public lecture by Zbigniew Szawarski, Professor of Philosophy, ethics specialist, Secretary to the Committee of Ethics in Science at the Presidium of the Polish Academy of the Sciences (January 2004).
(3) Installation of an internet forum to encourage public discussion.
(4) Four round-table discussions with representatives of the Catholic Church, other religions, the sciences and politics, who have a "liberal" stance on the problematic and a concluding meeting with representatives of all discussions.
(5) Constant contact with the mass media.
(6) Analysis of the expression of opinion on the internet forum.
(7) Checking and, if applicable, revision of the statement formulated by the government (13 January 2004) by the end of April 2004.

This schedule was not entirely kept to, as the concluding meeting took place in June and the analysis of the opinions expressed in the online forum was never published on the ministry's website. The title of the consultation program, Life for Life, aims to frame the issue within the pro-life discourse, thus facilitating its accord with the Catholic side. In this context, research on embryos is presented as a problem of deciding between two conflicting public goods regarding the protection of life: embryos and people with illnesses. The question raised is: would it be justifiable to take the life of an embryo for the sake of saving another life by developing new methods of treatment for incurable illnesses through research on embryos? What would speak in favor of this argument, as the government and other advocates of the research have pointed out, is that only spare embryos from IVF treatment are used, which are clearly destined to "die" in the first place. Advocates have also referred to the Polish law on transplantations by construing the use of spare fertilized egg cells from IVF as analogous to donating organs by persons who are brain dead (MNI 2003: 18, 21, 30; MNI 2004b: 8, c: 7, 17). During the round-table discussions, Catholic representatives disapproved of the use of such analogies. According to Wojciech Bołoz, priest, professor and head of the Center for Bioethics at the Cardinal Stefan Wyszyński University, the difference is that organ donors are deceased and that embryos are still alive (MNI 2003: 23).

Despite the fact that the deliberations showed that the Church has no uniform position on the topic and that attempts to find a compromise are respected, no actual proposal toward finding an agreement was ever accepted. One such approach to

[16]Available at <http://kbn.icm.edu.pl,komorki_macierzyste,20040126_2.html>, last accessed: 24 June 2004.

find a compromise is based on a technique that extracts stem cell lines without destroying the embryo stem cell. Another suggestion was based on the idea that only "suboptimal" fertilized egg cells are to be used, that is, egg cells that will clearly never be able to fully develop into a human even if implanted into a uterus (MNI 2003: 22). The former suggestion was dismissed straight away, because it cannot guarantee that the egg cell will not be "killed". The latter, although it may sound life protecting, was deemed unacceptable because it requires the recognition of protection based on phases of life development and on a principle of selection based on the criterion of the cell's ability to live.

Stem-cell research opponents, regardless of whether they are "independents" or church representatives, support two principles. One is an axiomatic statement according to which human dignity and human rights are indivisible. The right to life is valid from the "first moment of life". Therefore, protection must not be based on phases of life development, as human rights are always valid, regardless of whether or not certain characteristics are developed. The second is a historical-sociological argument, which claims that budging from this axiom in any direction would automatically lead down a slippery slope. This is where a possible discrimination of people with disabilities comes into play, and the lines of this argument are mainly historical. Six million deaths during World War II, Nazi eugenics and the eugenic practices in Scandinavian countries are mentioned (MNI 2003: 23, 29, MNI 2004a: 10). A commentary by Andrzej Rzepliński, an outstanding legal expert (see footnote 2), illustrates the extent to which comparisons from the arsenal of anti-totalitarianism are off base. He asserts that an embryo is even more vulnerable than the demonstrators who stood face to face with the tanks on Tiananmen Square in Beijing (MNI 2003: 19–20). His argument peters out in resignation that "we" are not going to be able to stop this research, because if it is not "us" doing the research, it will be Russia, North Korea and China.

The discussants repeatedly distanced themselves from the research practices of other countries. It is quite conspicuous that the "east" – Russia and Ukraine – have been taken as the overwhelming example of bad practice due to their lack of restrictions (MNI 2003: 25, MNI 2004c: 21). The positions taken toward the politics of the "west" have been more contradictory. Supporters of stem cell research do not want to see Poland pushed into the role of a backward country. The opponents see an opportunity for Poland to make history by taking an opposing stance on the issue. This approach is embedded in a discourse around Poland's "absolutely unique" history, and argues that opposing this research could be a statement that would go down in history, as Piotr Cywiński, the president of the Catholic Intelligence Club put it (MNI 2004a: 9). Such comments are rooted in a narrative of the heroic history of Poland and express the desire to position Poland within a united Europe as a people of morals and admonition.

A central point in this discussion touches on the issue the stem-cell research opponents have tried to avoid, yet one which has proven to be unavoidable: when life begins and where legal protection steps in. The constitution and the Act on Family Planning have both determined that the protection of life starts at the moment of conception, yet have both neglected to further define this moment, as former judge of the Constitutional Tribunal, Tomasz Dybowski, noted during the round table (MNI 2003: 21; MNI 2004a: 12). Opponents of human stem-cell research, particularly church representatives, are characterized by a remarkable combination of "traditional" and "modern" language use. Their semantics are particularly effective because they by refusing any further explication of "life" seem to present an unequivocal standpoint, whereas the research advocates only have ambiguities to offer.

The term "conception", which has become part of the language of the public and of legal terminology, has its origins in a historical era when the "fertilization" process inside a woman's body was unknown. "Conception" was, if it were not a divine gift, the result of a physical union between man and woman. Assisted reproduction techniques differentiate between sexuality and fertilization, as they create a "life" outside the body of the woman. The protection of life in the age of "conception" could thus only refer to a life to be born or to the pregnant woman. The whole idea of "conception" in the present age of assisted reproduction and genetics blurs the categorical differences between "life" inside a woman, "life" inside a Petri dish and a born child (Duden 1994, Wiesemann 2003). It is precisely this lack of explicitness that the advocates for research seek to question, e.g. with stories on moral dilemmas (to save one child or 50 000 spare embryos) or by highlighting misguided logical conclusions (MNI 2004c: 7/11f./14). Pro-lifers found their arguments on genetics. For them, conception is the union of the female egg cell with the male semen or the creation of a "unique genetic code" (MNI 2003: 29; MNI 2004a: 6). The State Secretary's objection that "human life is, in effect, impossible without planting a fertilized germ cell in a woman's womb" impelled a priest to respond by saying that it is not the location that is decisive in determining if something is alive or not (MNI 2003: 26, 28; MNI, 2004a: 3, 13).

The difference between the existential relationship between the woman and the embryo that develops inside her and the embryo inside a Petri dish is reduced here to merely being a difference in location. Pregnant women figure here, in a similar way to the abortion debate, as bearers and wombs. Pro-life discourse constructs a disembodied form of human reproduction. It does not consider the dependence of a human being on a woman for it to come into the world. Hence, it sounds ironic when pro-life advocates deem a human life synonymous with a technologically produced artifact, which is what one of the participants at the roundtable talks argues here (MNI 2004b: 12):

> Thanks to ultrasound, now everyone can see on screen what a human looks like during the first few weeks of its development. It is thus possible that, in the near future, science will advance to the point that it will become possible to witness the development of human life during its first hours or days. The discussion on the use of embryonic stem cells will then become invalid, because we will be able to see on the monitor that the embryo is a human being from the moment of its conception.

Again we meet here the remarkable way of reasoning, already illustrated earlier in this article, in right-wing discourse, in which facts are presented as proof of the truth of value statements. This argumentation implies an interesting conception of the democratic public sphere: science is expected to put an end to democratic debate.

The actual ethical conflict, between the bodily integrity of a woman and the protection of an embryo, is silenced. At the same time women are made unviable. At least some of the pro-lifers hope that science will make women obsolete in the processes of human procreation. A professor of cell biology and an internationally respected scholar from Krakow University devised a solution for spare embryos resulting from IVF treatment. Instead of "killing" embryos they could be saved in a frozen form for as long as possible in the hope that in the near future it will be technologically possible for them to develop into a human being without necessarily being implanted in a uterus (Korohoda 2002).

The consultation program did not provide the solution for a compromise the government had hoped to find. At the end of the project, the ministry highlighted the specific character of these round-table discussions. The novelty of this procedure, as the government officials stressed in their concluding words, was the fact that the talks had taken place at all, that they were recorded, and that these protocols were rendered publicly accessible on the internet. Government officials underscored the tolerance demonstrated toward one another's positions. They felt that this had made it possible to maintain different positions without condemning one another for their different approaches. Thus, they expressed hope that it had become possible for both sides to acknowledge one another's arguments, and that this discussion is now no longer the place to call embryo research advocates "child murderers" (MNI 2004c: 30–35). The State Secretary referred here to the language in the abortion debate and in articles on the consultation with the title "Death for Life" that appeared in the League of Polish Families' populist party newspaper (*Nasz Diennik*, 8 March 2004).

4.6 Conclusion and Outlook

The aim of this chapter was to account for Poland's puzzling policy pattern concerning human embryos, where very restrictive abortion legislation coexists with an unregulated hence permissive policy regime concerning the biomedical handling of human egg cells. I have argued that this is best understood from the perspective of a discursive institutionalist approach. In the Polish case the rules of political discourse and the institutional policy arrangements reinforce each other in creating an agonistic policy style which makes it extremely difficult to establish compromises on contentious issues. Consequently, this policy style favors polarized debates and fierce conflicts or strategies of evading the issue altogether. A central characteristic of Poland's democratic politics is that it evolved around cultural cleavages. The "cultural wars" which have dominated the political agenda ever since the fall of communism can be understood as a legacy inherited from the political struggles of the 1980s. This included unresolved tensions between different visions of citizenship as well as the oppositionary discursive framework based on moral categories of human dignity, whose irrefutable vigor drew more on transcendent catholic values than on secular human rights.

As I have shown in this article, the struggle over abortion was strongly marked by this conflict and shaped the cultural terrain for the articulation of Poland's new political and social order. The restrictive abortion legislation became a kind of founding compromise of the democratic Poland, on behalf of women's rights and bodily integrity. When the leftist government came to power in 2001, it refrained from a revision of the restrictive abortion law. Despite the fact that the Polish citizenry was divided on the abortion issue, a strongly value laden reframing of "life" as the highest good in democracy became hegemonic and established a discourse of "public morals", which makes it very difficult to debate the ethical dilemmas and legal challenges that new technologies pose. Ironically pro-lifers use genetic knowledge to support their normative claims as "natural" hence undisputable.

In the aftermath of the "abortion war" of the 1990s a silent compromise ruled: assisted reproductive technologies and the handling of human egg cells became a political non-issue. It seemed as if neither of the political camps wanted to politicize the topic, as they feared it could give rise to a new "value war". This does not mean that the new technologies were not publicly debated. However within discussions on cloning and stem cells domestic legal regulations had a remarkably low profile in Poland. This was also the case in the "societal consultation" on financing embryonic stem cell research which was under scrutiny in this chapter.

The consultation was an exercise in public involvement recommended by the European Commission. The novelty of this procedure was the fact that the talks had taken place at all and the rivals in the "abortion war" came together at the same table. The consultation exercise itself demonstrates the value of such procedures and particularly reveals their potential for facilitating communication on controversial policy issues, despite the antagonistic political climate. It confirms experiences with civic dialogues on other policy issues in Poland which have taken place as part of EU's recommended "New Mode of Governance" (Kulawik 2009b). However the consultation did not help to overcome the politics of non-decision.

Again it was the EU that required the Polish government to act; more precisely the EU Tissues and Cells Directives of 2004 and 2006 which imposed on every member State the need to establish a legal framework in order to ensure high standards of quality and safety with respect to the procurement, testing, processing, storage and distribution of tissues and cells (EU Directive 2004, 2006). The directives prescribed, among other things, that all persons and establishments who dealt with tissues and cells had to be accredited and licensed. These directives implied that an unregulated application of assisted reproductive technologies, prevalent in Poland, was not able to continue. The member States should comply with the directive September 2007, at the latest. Poland has up to now – April 2010 – not passed the required regulations. The leftist government started drafting a bill in 2005 which was never finalized. During the following 2 years of the populist coalition government led by the *Law and Justice Party* no attempt to comply with the EU regulations was made (Pezda 2008).

Only when the center-right *Civic Platform* came to power in autumn 2007 was a legislative process initiated. When the Minister of Health, Ewa Kowacz, in November 2007 announced the planned crafting of a law regulating "In Vitro", as the issue is referred to in Polish, she also mentioned the consideration of financing fertilization treatment with public funding. The plans triggered a lively and up to now, ongoing debate. A statement of the Polish Episcopat set the tone. Despite the fact that the Catholic Church for years tolerated the silent compromise, namely the coexistence of restrictive abortion legislation with an unregulated (and for women more risky) assisted procreation, its stand on the planned regulations, did not come as a surprise. The pragmatism – or what might be called hypocrisy – towards the "hidden" and "unofficial" vanished in the light of the public and transformed into dogma and hate speech. In a letter directed to the Polish Parliament the Episcopat's Family Council called "In Vitro" a "sophisticated way of abortion" in which numerous embryos are destined to "die". What is suggested here is that medical doctors as well as women and men who are involved in assisted reproductive techniques are in fact "child murderers". The letter condemns assisted fertilisation also with the argument, that the birth of children is gained through the "death of their brothers and sisters". This is an infamous claim as it declares that children who are brought into the world through biomedical techniques are born with a guilt on behalf of other "children" (Episkopat Polski 2007).

The language of the church does not allow for any distinction between born children, fetuses and egg cells. It lacks any tolerance for other value systems and continues a predemocratic tradition. The controversy about "In Vitro" became a constitutive conflict about the boundaries between the state and the church as well as the public and the private. The Prime Minister urged for a "human debate" not a war. Donald Tusk staked out three focal points for the legislation: the method should be regulated, legal and accessible even to those who lack resources, hence at least to some extent covered by public funding (Czaczkowska 2008). Since 2008 a continuous series of conferences, hearings and expert panels deliberating the issue took place. An expert commission, appointed by the Prime Minister, was endowed with the task to draft a law proposal concerning bioethical issues, including assisted procreation, embryo research and a bioethical council. The final report presented in October 2008 offered no clear legislative recommendations but two quite different versions of regulations, diverging on three focal points: access limited to married or open to cohabiting couples, application of preimplementation genetic diagnosis and admission of the freezing e.g. cryopreservation of so called spare embryos (Kancleria Prezesa Rady Ministrów 2008).

A major problem of the policy making process is, that the governing party, the *Civic Platform* is divided on the issue. Attempts to accomplish a compromise within the parliamentary party group failed (*PO nie złoży projektu ustawy o* in vitro 2009). The most controversial topic proved to be the freezing of so called "spare embryos". To be sure: the hormone stimulation required for a fertilization in vitro frequently creates more egg cells than can be implanted during one so called "fertilization cycle". The preservation of spare eggs cells allows several fertilization cycles to be carried out (if needed), without repeating the risky ova retrieval procedure. The internationally common practice is to preserve fertilized egg cells because oocytes solely did usually not endure the freezing process. Even Germany with it's quite restrictive regulations allows the freezing of fertilized egg cells, though before the nuclear fusion between egg-cell and semen took place. According to the legal definition fertilized egg cells in the pronuclear stage are not regarded as "embryos".

In the Polish context, where fertilized egg cells are referred to as "conceived children" the preservation issue has become extremely laden. Again historical analogies are employed which are off base. Jaroslaw Gawin, the head of the expert group appointed by the Prime Minister, created an analogy between cryo preservation in which fertilized egg cells are placed in liquid nitrogen, e.g. put in a gas and "gassing", the mass murder in Nazi concentration camps (*Kończą się prace nad rekomendacjami ws. ustawy bioetycznej,* 2009). This illustrates the loss of distinction ability and the brutalization of public language in Poland, where "Life" and "History" are instrumenalized in political battles. We should not forget women's bodies here. An eventual ban on cryo preservation results in much higher health risks for women. Again the protection of "Life" figures higher than women's health and bodily integrity.

In autumn 2009 three law proposals have been submitted to the Polish Parliament. The bill by the right-wing *Law and Justice Party* demands the prohibition of assisted reproduction techniques. Members from the governing party *Civic Platform* have submitted two rather divergent bills which parallel the two proposals of the commissions report. Up to now the parliamentary consultation process is not finished. If I may speculate here about the future legislative framework regarding the biomedical handling of human egg cells in Poland, which has to be passed because EU directives do not allow the continuance of the old unregulated condition: My hypothesis is that Poland will embark on a similar path to that followed by Italy (Metzler 2007). Italy has moved from an unregulated permissive policy design to the most restrictive regime in Europe. The supportive evidence is the discursive shift since the issue was put on the political agenda. The public debate of the former non-issue was kicked off by the aim to make biomedical practices both safer and more accessible through public oversight and funding. The debate is instead dominated by the prohibition frame.

References

Abels, G., K. Braun, and T. Kulawik. eds. 2003. Geschlecht und Biomedizinpolitik. Vergleichende Perspektiven. Special Issue: *Österreichische Zeitschrift für Politikwissenschaft* 32:3.

Bacchi, C., and Ch. Beasley. 2002. Citizen bodies: Is embodied citizenship a contradiction in terms? *Critical Social Policy* 22 (2):324–352.

Bleiklie, I., M., Goggin, and C., Rothmayr. 2004. *Governing assisted reproductive technology. A cross-country comparison*. London, New York: Routledge.

Bołoz, Wojciech, and Gerhard Höver. 2002. *Utylitaryzm w bioetyce.Jego założenia i skutki na przykładzie poglądów Petera Singera*. Warszawa: Publikacje Centrum Ekologii i Bioetyki Kardinał Stefan-Wyszyński Universytet. German edition =Utilitarismus in der Bioetik. Münster–Hamburg–London: Litverlag.

Brier, R. 2009. The roots of the 'fourth republic': Solidarity's cultural legacy to Polish politics. *East European Politics and Societies* 23(1):63–85.

Chyrowicz, Barbara (ed.). 1999. *Klonowanie człowieka. Fantazje, zagrożenia, nadzieje*. Lublin: Towarzystwo Naukowe Katolickiego Uniwersytetu Lubelskiego.

Chyrowicz, B. 2000. *Bioetyka i ryzyko. Argument 'równi pochyłej' w dyskusji wokół osiągnięć współczesnej genetyki*. Lublin: Towarzystwo Naukowe Katolickiego Uniwersytetu Lubelskiego.

Czaczkowska, E. 2008. Tusk: wszyscy powinni mieć dostęp do in vitro. *Rzeczpospoplita*, 27 November, available at http://www.rp.pl/artykul/81429,225384-Tusk--wszyscy-powinni-miec-dostep-do-in-vitro-.html.

Duden, B. 1994. *Der Frauenleib als öffentlicher Ort. Vom Mißbrauch des Begriffs Leben*. München: Luchterhand.

Döring, M., and J., Zinken. 2005. The cultural crafting of embryonic stem cells: the metaphorical schematisation of stem cell research in the Polish and French press. Available at www.metaphorik.de. Accessed 26 April 2006.

Elam, M., and M., Bertilsson. 2003. Consuming, engaging and confronting science: The emerging dimension of scientific citizenship. *European Journal of Social Theory* 6: 233–251.

Episkopat Polski. 2007. Rada do spraw rodziny, http://www.episkopat.pl/?a=dokumenty\penalty\z@KEP&doc=20071218.

EU Directive 2004/23/EC of 31 March 2004 on setting standards of quality and safety for the donation, procurement, testing, processing, preservation, storage and distribution of human tissues and cells http://eur-lex.europa.eu/LexUriServ/LexUriServ.do?uri=OJ:L:2004:102:0048:0058:EN:PDF.

EU Directive 2006/86/EC of 24 October 2006 implementing Directive 2004/23/EC as regards implementing Directive 2004/23/EC of the European Parliament and of the Council as regards traceability requirements, notification of serious adverse reactions and events and certain technical requirements for the coding, processing, preservation, storage and distribution of human tissues and cells http://eur-lex.europa.eu/LexUriServ/site/en/oj/2006/l_294/l_29420061025en00320050.pdf.

Fuchs, G. 2003. *Die Zivilgesellschaft mitgestalten. Frauenorganisationen im polnischen Demokratieprozess*. Frankfurt/M.: Campus Verlag.

Gal, S. and G. Kligman. eds. 2000. *Reproducing gender. politics, publics, and everyday life after socialism*. Princeton, NJ: Princeton University Press.

Gaskell, G., et al. 2006. Europeans and Biotechnology in 2005: Patterns and Trends. A report to the European Commission's Directorate-General for Research, Available at http://www.ec.europa.eu/research/press/2006/pdf/pr1906_eb_64_3_final_report-may2006_en.pdf. Accessed 18 July 2007.

Gazeta Wyborcza. 2004. Gen ciekawości. In *Z wybitnymi naukowcami rozmawiają dziennikarze Gazety Wyborczej*. Warszawa: Prószyński i S-ka.

Gottweis, H. 1998. *Governing molecules. The discursive politics of genetic engineering in Europe and in the United States*. Cambridge, MA: The MIT Press.

Graff, A. 2009. Gender, sexuality, and nation – here and now. reflections on the gendered and sexualized aspects of contemporary Polish nationalism. In *Intimate Citizenships. gender, sexualities, polities*, ed. E. Oleksy. New York, London: Routledge.

Górski, E. 2002. From 'socialist' to postmodern pluralism in Poland. *East European Politics and Societies* 16(1):249–277.

Hierlemann, D. 2005. *Lobbying der katholischen Kirche. Das Einflussnetz des Klerus in Polen*, Vol. 211. Wiesbaden: Verlag für Sozialwissenschaften.

Holmes, S. 1996. Cultural legacies or state collapse? Probing the postcommunist dilemma. In *Postcommunism: Four perspectives*, ed. M. Mandelbaum, 22–76. New York, NY: Council of Foreign Relations.

Immergut, E.M. 1998. The theoretical core of the new institutionalism. *Politics and Society* 26(1):5–34.

Jacorzyński, W., and M., Kozłowski. 2005. Auf dem Weg nach (n)irgendwo: die polnische Debatte zur Stamzellforschung. In *Grenzüberschreitungen: kulturelle, religiöse und politische Differenzen im Kontext der Stammzellenforschung*, eds. W. Bender, Ch. Hauskeller and A. Manzei, 469–487. Münster: Agenda-Verlag.

Jarosław, M., and W., Wróbel. 1993. Prawnokarna ochrona dziecka poczętego. *Państwo i Prawo* 5:35–44.

Jasanoff, S. 2005. *Designs on nature. science and democracy in Europe and the United States*. Princeton, NJ: Princeton University Press.

Just, E.M. 2008. New reproductive assemblages: Understanding, managing and 'using' human in vitro fertilization (IVF). PhD Dissertation, Utrecht University.

Kaluza, A. 1998. *Der polnische Parteistaat und seine politischen Gegener 1944–1956*. Frankfurt/M: Vittorio Klostermann.

Kancleria Prezesa Rady Ministrów. 2008. Raport Zespoła ds. Konwencji Bioetycznej, 28 october 2008, Warsaw.

Katolo, A.J. 2000. *Embrion ludzki, osoba czy rzecz? status i moralne prawa nienarodzonego we włoskiej literaturze teologiczno – bioetycznej po Soborze Watykańskim II*. Lublin, Sandomierz: Wydawnictwo Diecezjalne w Sandomierzu.

KBN 2003. Stanowisko Komitetu Badań Naukowych w sprawie badań nad komórkami macierzystymi i wykorzystaniem ich w medycynie (18 September 2003). Available at <http://kbn.icm.edu.pl/komorki_macierzyste/20040126_1.html>, last accessed: 24 June 2004.

Komitet Etyki w Nauce przy Prezydium PAN. 1997. Zagrożenia i mity wynikające z rozwoju genetyki. Odbitki materiałów ze wspólnej konferencji... 23 October 1996. *Special Issue: Nauka* 3(2).

Korohoda, W. 2002. Inżynieria komórkowa i tkankowa na początku XXI wieku – nowe nadzieje i nowe zagrożenia. In *Zagrożenia Cywilizacyjne*, 123–137, Vol 5. Kraków: Polska Akademia Umiejętności.

Kończą się prace nad rekomendacjami ws. ustawy bioetycznej, Rzeczpospolita, 4 March 2009, Availabe at http://www.rp.pl/artykul/81429,271685_Koncza_sie_prace_nad_rekomendacjami_ws__ustawy_bioetycznej.html, last accessed 6e april 2009.

Kramer, A.-M. 2009. The Polish Parliament and the making of politics through abortion: Nation, gender and democracy in the 1996 Liberalization Amendment Debate. *International Feminist Journal of Politics* 11 (1): 81–101 March.

Kubik, J. 2003. Cultural legacies of state socialism. history making and cultural-political entrepreneurship in postcommunist Poland and Russia. In *Capitalism and democracy in Central and Eastern Europe: Assessing the legacy of communist rule*, eds. G. Ekiert and S.E. Hanson, 317–352. Cambridge: Cambridge University Press.

Kulawik, T. 2009a. Feminist discursive institutionalism. *Gender and Politics*, Special Issue "Critical Perspectives on Feminist Institutionalism." 5(2):262–271.

Kulawik, T. 2009b. Science Policy and Public Accountability in Poland. The case of embryonic stem cell research. *Science and Public Policy* 36(7):469–482, July.

Łuków, P. 2001. Moralne aspekty klonowania człowieka a światopglądy. *Medycyna wieku rozwojowego* 5(suppl 1):103–111.

MNI, Ministry of Science and Information Technology. 2003. Ministerstwo Nauki i Informatyzacji. Komórki macierzyste – życie za życie? Wykorzystywanie zarodkowych komórekmacierzystych – aspekty medyczne i etyczne (15 December).Available at http://kbn.icm.edu.pl/komorki_macierzyste/20040217.html. Accessed 24 June 2004.

MNI. 2004a. Ministerstwo Nauki i Informatyzacji. Komórki macierzyste – życie za życie? Wykorzystywanie zarodkowych komórek macierzystych – aspekty medyczne i etyczne (8 March). Available at http://kbn.icm.edu.pl/komorki_macierzyste/20040414.html. Accessed 24 June 2004.

MNI. 2004b. Ministerstwo Nauki i Informatyzacji. Komórki macierzyste – życie za życie? Wykorzystywanie zarodkowych komórek macierzystych – aspekty medyczne i etyczne (19 April). Availabe at http://www.mnii.gov.pl/_gAllery/30/95/3095.rtf. Accessed 24 June 2004.

MNI. 2004c. Ministerstwo Nauki i Informatyzacji, Komórki macierzyste – życie za życie? Wykorzystywanie zarodkowych komórek macierzystych – aspekty medyczne i etyczne (23 June). Available at http://kbn.icm.edu.pl/komorki_macierzyste/. Accessed 15 September 2004.

MNI. 2005. Ministerstwo Nauki i Informatyzacji, Komórki macierzyste – życie za życie? Podsumowanie Debaty. Available at <http://www.mnii.gov.pl/mnii/index.jsp?place=Lead02&news_cat_id=135&news_id=695&layout=1&page=text>, last accessed 11 August 2005.

Matuchniak-Krasuska, A. 1991. Kategorie i reguły polskiego dyskursu o aborcji. In *Cudze problemy. O ważność tego, co nieważne. Analiza dyskursu publicznego w Polsce,* eds. M. Czyżewski, K. Dunin, and A. Piotrowski, 100–122. Warszawa: Ośrodek Badań Spotecznych.

Medycyna wieku rozwojowego. 1999. Special Issue on Cloning 3(3), supplement. *Medycyna wieku rozwojowego* 2001. Special Issue on Cloning 5(1), supplement.

Metzler, I. 2007. 'Nationalizing embryos': the politics of human embryonic stem cell research in Italy. *BioSocieties* 2(4):413–427.

Michalski, K. 2003. Technikbewertung in Polen. Tradition, Wandel, Perspektiven. In *Technik-System-Verantwortung*, ed. K. Kornwachs, 425–434. Münster: Lit Verlag.

Nauka. 2003. "Nauki przyrodnicze na drodze pomiędzy bioetyką i biopolityką. Materiały z konferencji. . .20.5. 2002." *Special Issue: Nauka* 10.

Pawlik, W. 1991. Spór o aborcje czyli sztuka parlamenternej erystyki. In *Cudze problemy. O ważność tego, co nieważne. Analiza dyskursu publicznego w Polsce*, eds. Marek Czyżewski, Kinga Dunin, and Anrzej Piotrowski, 123–151. Warszawa: Ośrodek Badań Spotecznych.

Pezda, A. 2008. In vitro w szufladzie. *Gazeta Wyborcza*, 6th march. Available at http://wyborcza.pl/1,87495,4814208.html

Planert, U. 2000. Der dreifache Körper des Volkes: Sexualität, Biopolitik und die Wissenschaften vom Leben. *Geschichte und Gesellschaft* 26(H. 4, S.):539–576.

PO nie złoży projektu ustawy o in vitro. 2009. *Rzeczpospolita*, 18.3. 2009, Available at http://www.rp.pl/artykul/81429,290762_Ustawa_o_in_vitro__odlozona_do_szuflady.html. Accessed 6 Apr 2009.

Przestalski, A., B., Suchocki, and T., Twardowski. 2001. Poland: Arrival of the Gene Law. In *Biotechnology 1996–2000, the years of controversy,* eds. G. Gaskell and M.W. Bauer. London: Science Museum.

Przyłuska-Fiszer, A. 2005. Die bioethische Diskussion über Reproduktionstechnologien und Embryonenforschung in Polen. In Hofmann, 55–84; Special Issue on Cloning, *Medycyna wieku rozwojowego*. 3, 3, suppl. (1999) and 5, 1, suppl. (2001).

Raciborski, J., and J.J., Wiatr. 2005. *Demokratie in Polen. Elemente des politischen Systems.* Opladen: Verlag Barbara Budrich.

Russel, A., and J., Vogler. eds. 2000. *The international politics of biotechnology. Investigating global futures.* Manchester University Press.

Safjan, M. 1992. O metodach rozwiązywania dylematów bioetyki. *Państwo i Prawo* 5:51–59.

Safjan, M. 2000. Nowy wspaniałym człowiek, Interview with Ewa Siedlecka and Sławomir Zagórski. *Gazeta Wyborcza*, Świąteczna (weekend supplement), 15–16 April, 10.

Safjan, M. 2001. Rozwój nauk biomedychnych a granice ochrony prawnej. In *Współczesne problemy bioetyki w obszare regulacji prawnych. Materiały z konferencji. . . .*, ed. Senat R.P., 29–54, 45–46. Warszawa: Dział Wydawniczy Kancelarii Senatu, 3 April 2001.

Safjan, M. 2003. Konwencja Bioetyczna – jej znaczenie dla kształtowania standardu prawnego. In *Zastosowanie ideologii w Medycynie A Godność Osoby Ludzkiej – aspekty etyczne i prawne,* ed. Polskie Towarzystwo Genetyki Człowieka, 28–42, 40. Warszawa: Wydawnictwo Naukowe PWN.

Safjan, M. 2004. Interview of the author with Marek Safjan, Constitutional Tribunal, Warsaw (1st July 2004).

Sandor, J. ed. 2003. *Society and genetic information: Codes and laws in the Genetic Era.* Budapest: Central European University.

Schmidt, V. 2008. Discursive institutionalism: The explanatory power of ideas and discourses. *Annual Review of Political Science* 11:303–326, June.

Staskiewicz, W. 2008. The legislative process. In *Democracy in Poland 2005–2007*, eds. L. Kolarska-Bobinska, J. Kucharczyk, and J. Zbieranek, 31–38. Warszawa: Instytut Spraw Publicznych.

Środa, M. 2005. Hindernisse und Chancen der Moraldebatte in Polen und dieMarginalisierung der Frauen. In *Biopolitik grenzenlos. Stimmen aus Polen*, ed. H. Hofmann, 85–106. Herbolzheim: Centaurus Verlag.

Twardowski, T., and A., Michalska. 2000. *Dylemmaty Współczesnej Biotechnologii z perspektywy biotechnologa iprawnika*. Toruń: Wydawnictwo Dom Organizatora.

UN, United Nations General Assembly. 2005. A/59/PV.82, 59th session, 82nd plenary meeting, (8 March): 7 (Mr Zyman).Available at http://www.un.org/law/cloning/. Accessed 18 July 2006.

Wiesemann, C. 2003. Wie kann über den Embryo in einer lebensweltlich angemessenen Weise gesprochen werden? Eine Kritik der Debatte um den moralischen Status des Embryo. In *Verkörperte Technik – Entkörperte Frau. Biopolitik und Geschlecht*, eds. S. Graumann and I. Schneider, 141–151. Frankfurt/M: Campus Verlag.

Youval-Davis, N. 1998. *Gender and Nation*. London: Sage.

Zielińska, E. (1995) "Z problematyki wykładni przepisów karnych dotyczących ochrony płodu ludzkiego. *Państwo i Prawo* (2):25–29.

Zielińska, E. 1999. Rady Bioetyczne i ich rola w rozwiązywaniu dylematów bioetyki. *Prawo i Medycyna* 1(2):28–37.

Zielińska, E. 2000. Between ideology, politics, and common sense: The discourse of reproductive rights in Poland. In *Reproducing gender. politics, publics, and everyday life after socialism*, eds. S. Gal and G. Kligman, 23–57. Princeton, NJ: Princeton University Press.

Zielińska, E. 2005. Der Schutz des Embryos und die Rechte der Frauen bei der Reproduktion in Polen. In *Biopolitik grenzenlos. Stimmen aus Polen*, ed. H. Hofmann, 216, 21–31. Herbolzheim: Vcentaurus Verlag.

Ziemer, K., and C., Matthes. 2004. Das politische system polens. In *Die Politischen Systeme Osteuropas*, ed. W. Ismayr aktualisierte und überarbeitete Auflage, Vol. 2, 189–246. Opladen: Leske+Budrich.

Chapter 5
Legal Ambiguities Concerning Medical Genetics in Poland – Searching For a Common Ground

Atina Krajewska

5.1 Introduction

Since the discovery of the double-helix structure of DNA several paradigm shifts have taken place in life and social sciences' discourse surrounding issues related to genetics.[1] The Human Genome Project (HGP), initiated in 1990, changed the focus of the debate from molecular-biology-based-*genetics*, which concentrated on the role and function of a single gene, to *genomics* preoccupied with processes between and inside genomes.[2] Paradoxically, however, the HGP initiated yet another change of perspectives, towards the so called *post-genomics,* which focuses on the interactions and dynamics between genetic and environmental components. "While [genetics] still tries to assimilate new and spectacular findings of genomic complexity to the concept of a particular gene with a more or less discrete physical structure defined by its boundaries as much as by its functional product(s), [post-genomics] accommodates its concepts to a molecular reality based on flexible entities that are defined by spatial organisation and location, and by sensitivity to intra- and extracellular signals" (Stotz et al. 2006). New emerging fields, including proteomics,[3]

The final version of this chapter was submitted to the editors in December 2009.

[1]Genetics, defined broadly, is the science of heredity and variation in living organisms. See: Griffiths et al. (2009).

[2]The Human Genome Project (HGP) refers to the international 13-year effort, formally begun in October 1990 and completed in 2003, to discover all the estimated 20,000–25,000 human genes and make them accessible for further biological study. Another project goal was to determine the complete sequence of the 3 billion DNA subunits (bases in the human genome). See: Human Genome Project Information, available at: http://www.ornl.gov/sci/techresources/Human_Genome/home.shtml. Retrieved on 3 May 2009.

[3]The study of the proteome, the complete set of proteins expressed by a genome, cell, tissue or organism, using the technologies of large-scale protein separation and identification. The term proteomics was coined in 1994 by Marc Wilkins from the UNSW School of Biotechnology and Bimolecular Sciences. Retrieved June 30, 2008 http://www.babs.unsw.edu.au/staff_directory/wilkinsm.html.

A. Krajewska (✉)
University of Exeter, Exeter, UK
e-mail: a.krajewska@exeter.ac.uk

P.T. Robbins, F. Huzair (eds.), *Exploring Central and Eastern Europe's Biotechnology Landscape*, The International Library of Ethics, Law and Technology 9,
DOI 10.1007/978-90-481-9784-2_5,

metabolomics,[4] and systems biology – the latter taking into consideration new factors, such as context, time and space (Ahn et al. 2006) – attract more and more attention.

The post-genomic era[5] is, thus marked by the departure from genetic determinism, essentialism and exceptionalism. It seems increasingly difficult to treat genetic information as the essential determinant of human existence, and a particularly sensitive class of personal data.[6] The highly predictive power of genetic testing in medical practice is limited to diagnosis of monogenic disorders, i.e. diseases inherited by a single pair of genes. Researchers point out that 'the belief that a few genes held the key to ridding the world of conditions such as cancer and diabetes has proved to be "plain wrong"'.[7] Thus, genetics and genomics appear not to have fulfilled many of their promises. These difficulties translate into major conceptual quandaries about the regulation of the use of DNA and biological data.[8]

Of course, this is not to say that life sciences play no role in our every day life. New biotechnological devices and genetic diagnostic tools have been extensively used in medicine for years. New branches such as medical genetics and medical genomics have developed.[9] Diagnostic (and therapeutic) methods coupled with assisted reproductive technologies (ART)[10] offer powerful tools of birth control and small scale or even wide-spread human enhancement. Even if we are not able to determine exactly to what extent a particular gene increases the risk of developing cancer, we may still want to employ pre-implantation genetic diagnosis in order to minimise this risk in our (future) children. Ethical objections that are routinely voiced against the use of genetics and reproductive technologies are often rightly said to lack robustness and persuasiveness.[11]

Still, one needs to be aware that, as predicted by critics in the early 1990s, 'the era when scientists decided about the direction of their research seems to be history. Scientists are now obliged to respond to calls for tenders issued by governments, the European Union and large companies following economic and

[4]Metabolomics is the global analysis of metabolites, small molecules generated in the process of metabolism. See also: Jenkins et al. (2004).

[5]Postgenomic era, 'in which genetic information will have to be examined in multiple health care situations throughout the lives of individuals' (Peltonen and McKusick 2001).

[6]The post-genomic paradigm emphasises the arbitrariness of taking genes as the most important casual factors in the development and functioning of organisms. Consequently, it highlights that the concept of the gene itself is a highly problematic conceptual and epistemological conundrum. I discuss this problem in legal terms in: Krajewska (2009a).

[7]'Professor Steve Jones in: Alleyne and K. Devlin (2009).

[8]I have analysed this problem in: Krajewska (2009b).

[9]The former is usually defined as 'a branch of biomedical science that studies the relationship between genes and health, and searches for an unknown gene that may be involved in a disease', whereas the latter attempts to translate effectively genome-based knowledge for the benefit of health care. In: Connor and Ferguson-Smith (1997).

[10]This term encompasses different methods used to achieve pregnancy by artificial or partially artificial means.

[11]For further discussion see: Harris (2007) and Somsen (2008).

business-strategic criteria' Gibbons et al. (1994). As the alliance between scientific knowledge, economics and politics becomes prominent, new expert elites emerge. This raises concerns about core democratic values such as citizens' participation and governmental accountability.[12]

Therefore, what becomes particularly salient in the post-genomic era is the criticism of the mere process of knowledge production. It has been pointed out that: 'the metaphorical language used by scientists to constitute their theories tends to be forgotten (. . .) and "genes", "genetic programmes" and "genetic codes" have come to be conceived as pre-discursive, empirical realities' (Rouvroy 2008). The 'geneticisation' process should be treated with great caution, as genetic knowledge production is seen as subjected to and determined by the (neo)liberal principle.[13] 'Intensive funding of genetic science should thus be seen less as an investment made for the benefit of future generations than as an interdisciplinary contribution to a new biopolitics' (Rouvroy 2008: 21).

And yet, the shift from genetics to post-genomics has hardly been acknowledged outside of laboratories. In public opinion, mainstream law and bioethics, the genocentric dogma remains almost unquestioned (Rouvroy 2008: 35). Consequently, legal provisions regulating genetic medicine appear at the national, supranational, and international level.[14] Although most international documents are not legally binding, they serve as an important point of reference for the national legislator. At the domestic level regulatory responses focus on genetic testing – postnatal, prenatal (PD), and pre-implantation genetic diagnosis (PGD) – and genetic biobanks (where biological samples are collected and stored).[15] As law usually fails to keep pace with new phenomena, legislators at all levels are slow in reacting to the recent paradigm shift towards post-genomics.

The launch of the Human Genome Project in 1990 coincided with the historic events after the 'Round Table' and the subsequent fall of the Iron Curtain. In the

---

[12]This problem has been thoroughly analysed by S. Jassanoff (2007).

[13]M. Foucault analyses liberalism not 'as a theory or an ideology – and even less, certainly, as a way for "society" to "represent itself. . ." – but, rather, as a practice (. . .) as a principle and a method of rationalizing the exercise of government, a rationalization that obeys – and this its specificity – the internal rule of maximum economy.', Foucault (1997).

[14]UN, UNESCO, WHO, HUGO, WMA, OECD and Council of Europe have issued many documents concerning, for instance human genome (UNESCO Declaration on Human Genome and Human Rights 1997), genetic data (UNESCO Declaration on Human Genetic Data 2003), genetic testing (Additional Protocol to the Convention on Human Rights and Biomedicine on Genetic Testing for Health Purposes 2008) and biobanks (OECD Guidelines for Human Biobanks and Genetic Research Databases 2009).

[15]In Germany the Act on Genetic Diagnosis has been adopted by the German Bundestag on the 23 April 2009. See: 'Gen-Diagnostik Gesetz verabshidet' in: Focus, 24 April 2009, available at: http://www.focus.de/gesundheit/ticker/recht-gen-diagnostik-gesetz-verabschiedet_aid_393030.html. Retrieved 1 May 2009, In Switzerland the Federal Law on the Genetic Testing of Humans (Loi fédérale sur l'analyse génétique humaine (LAGH) FF 2004 5145), came into force in 2006. In the USA the famous Genetic Information Non-Discrimination Act (GINA Pub.L. 110–233, 122 Sta. 881) has been enacted on 21 May 2008. For regulation of biobanks in Iceland, Estonia, Hungary, and the UK see: Maschke (2005).

20 years that followed, Central and Eastern European states were preoccupied with democratisation and transformation processes. Therefore, relatively little is known about the legal discourse surrounding life sciences in these countries. Poland is no exception to this rule. The country is perceived as extremely conservative with restrictive legislation, influenced by the Catholic Church that dominates the ethical, political, and legal discourse, and shapes government's policy. Polish members of the European Parliament (MEP) are famous for their Catholicism and shocking pro-life campaigns.[16] The famous Strasbourg case of *Alicja Tysiac v. Poland*[17] concerning the denial of pregnancy termination even within the boundaries of existing law confirms this view. On the other side of the political spectrum liberal and feminist movements are (often incorrectly) perceived as radical and uncritical towards scientific progress (Jacorzynski and Kozlowski 2005). The debate over bioethics is highly polarised and seems to be centred around issues of abortion and the moral and legal status of an embryo.

At the same time, however, Poland like all other Central European countries, adopted the principles of a free market economy such as competition, privatisation and deregulation. (Neo) liberalism became a new dogma of the transformed state. These changes have provided scientists and doctors with many opportunities to engage with international research. They have influenced the development of medical genetics and genomics. In recent years Poland has been witnessing a rapid increase in availability of genetic testing, not only for forensic and family law, but also for health related purposes. New population screening studies and research on gene functions and new diagnostic methods of genetic disorders have been launched; private biobanks of umbilical cord blood and bone marrow have been established. Private IVF clinics now offer genetic services, including neonatal, prenatal and preimplantation diagnosis.[18] The Catholic social thought engrained in Polish society is very disapproving of the liberal transformation that has been taking place recently. Still, politicians, media and the public do not seem to be interested in regulating medical genetics (unless it concerns human embryos). Legal initiatives undertaken to address issues arising in connection to the changes in life sciences may thus seem ambiguous and incoherent.

Therefore, it is important to understand, how Polish regulators, reconcile liberal principles of free market economy, with the Catholic social thought so prominent in Polish politics. Should Poland be seen as the *black sheep* of Europe? Is it really a country that hinders harmonization processes at European level? What are the strengths and the weaknesses of the Polish regulatory framework in light of the

---

[16]European Jewish Congress Website, available at: http://www.eurojewcong.org/ejc/news.php?id_article=159. Retrieved on 25 April 2009.

[17]European Court of Human Rights (ECtHR), *Tysiąc v. Poland*, 20.03.2007 r., Application No 5410/03.

[18]See e.g.: European Bank of Umbilical Cord Blood 'Motherhood', available at: http://www.macierzynstwo.pl/index.php. Retrieved on 20 May 2008; Stem Cell Bank 'Progenis' Ltd. Retrieved: http://www.progenis.pl/index.php/kontakt.html; Polish Stem Cell Bank S.A. (joint stock company), available at: http://www.pbkm.pl/. Retrieved on 20 May 2008.

recent paradigm shift that has been taking place in life sciences? To what extent is the Polish law fitted for the post-genomic era and the human rights standards constituting the basis of a democratic state? Where is the common ground upon which Europe could reach a much needed consensus?

The analysis will be divided into two main parts concerning: (a) medical research and (b) medical practice. It will focus on three main aspects of medical genetics, namely on genetic testing of individuals, population screening, and biobanks. Being aware of the conceptual difficulties of disentangling DNA as a chemical substance, from the genetic information it entails, the analysis will nevertheless reflect the distinction made by the legislator between: (a) data protection law and, (b) provisions concerning the use of biological material (human body parts).

5.2 International Legal Framework

Over the last 20 years Poland has joined important international organizations, including the Council of Europe in 1991, NATO in 1999 and the EU in 2004. This has brought changes to the Polish legal system, which before was influenced by three major legal and political traditions, i.e. French civil law, German/Austrian positivism and Marxism. Human rights standards became an integral part of the Polish constitution and a basic point of reference in the process of building a civic society. For many years Poland has been a signatory to most of the important human rights documents of a general nature including the Universal Declaration of Human Rights (UDHR 1948), and the two International Covenants on Civil and Political Rights; and Social, Economic and Cultural Rights (1966). Although these acts are not directly concerned with research or medical practice, they contain human rights that serve as directives for the interpretation of international and national medical law.

As far as international law and bioethics are concerned Poland participated in drafting the UNESCO Declarations on the Human Genome and Human Rights (1997), Human Genetic Data (2003), and Bioethics and Human Rights (2005).[19] The latter documents are not legally binding and their legal status has been disputed. However, they contain relatively detailed provisions concerning genetics and genomics and certainly constitute a point of reference for domestic legislators, ethics committees and researchers. At the same time, since there are no instruments of enforceability, the impact of these documents on research practice differs across the research and medical professions. The Polish National Chamber of Physicians and other medical associations are bound by the WMA Helsinki Declaration,[20] which has influenced Polish constitutional and medical law.

[19] All UNESCO documents on Bioethics are available at: http://portal.unesco.org/shs/en/ev.php-URL_ID=1372&URL_DO=DO_TOPIC&URL_SECTION=201.html. Retrieved on 29 April 2009.

[20] World Medical Association, Helsinki Declaration 1964, last version adopted by the 59th WMA General Assembly, Seoul, October 2008, available at: http://www.wma.net/e/policy/b3.htm. Retrieved on 1 May 2009.

At European level, following membership of the Council of Europe, Poland ratified the European Convention on Human Rights (1950) in 1993.[21] In 1999 the Government also signed the Convention on the Protection of Human Rights of the Human Being with Regard to the Application of Biology and Medicine, signed in Oviedo in 1997 (later also: the Convention on Human Rights and Biomedicine, Oviedo Convention or BioConvention) and the Additional Protocol on the Prohibition of Cloning of Human Beings (1998).[22] Since no ratification of these instruments has yet taken place, Poland is not bound by their provisions. Still, it has to refrain from acts which would defeat the object and purpose of the Treaty.[23] Moreover, the Convention is the only legally binding international treaty on biomedicine and bioethics in the world. It entails comprehensive rules on medical practice and research. Furthermore, in light of the latest recommendation issued by a government's ad hoc bioethics committee in October 2008, there is reason to believe that the ratification will take place in the near future.[24]

In accordance with Article 87 paragraph 1 of the Polish Constitution, once ratified the Convention will become a binding source of law in Poland. As stated in Article 91 paragraph 1 of the Constitution 'a ratified international agreement, upon its publication in the Journal of Laws of the Republic of Poland, constitutes a part of the national legal system and will be applied directly, unless its application requires a Regulation to be issued' (Constitution of the Republic of Poland 1997).[25] This means that, despite the fact that Poland favours a dualistic system of relations between international and national law, no special enacting legislation is necessary to give effect to the Convention's provisions.[26] Of course even in the absence of

---

[21] Council of Europe, Convention for the Protection of Human Rights and Fundamental Freedoms Rome, 4 November 1950, as amended by Protocol No. 11, available at: http://conventions.coe.int/treaty/en/Treaties/Html/005.htm. Retrieved on 30 October 2008.

[22] Council of Europe, Convention for the Protection of Human Rights and Dignity of the Human Being with regard to the Application of Biology and Medicine: Convention on Human Rights and Biomedicine, Oviedo, 4.IV.1997, available at: http://conventions.coe.int/Treaty/EN/Treaties/Html/164.htm. Retrieved on 28 October 2008.

[23] Art. 18 of the Vienna Convention on the Law of Treaties (1969) stipulates that: A State is obliged to refrain from acts which would defeat the object and purpose of a treaty when: (a) it has signed the treaty or has exchanged instruments constituting the treaty subject to ratification, acceptance or approval, until it shall have made its intention clear not to become a party to the treaty; or (b) it has expressed its consent to be bound by the treaty, pending the entry into force of the treaty and provided that such entry into force is not unduly delayed. See: United Nations, *Treaty Series*, vol. 1155, p.331.

[24] The Committee is expected to announce the results of its work at the beginning of October 2008. It will propose necessary amendments to the existing law and new statutes on in vitro fertilisation and other biomedical issues. See: Interview with Jarosław Gowin, the Chair of the Bioethics Committee, Radio TOKFM, Retrieved September 18, 2008, http://serwisy.gazeta.pl/tokfm/1,53880,5707747,Gowin__Szykuje_sie_bardzo_pracowita_jesien.html.

[25] Constitution of the Republic of Poland, 2 April 1997, Dz. U. Nr. 78 poz. 483.

[26] The relationship between international and domestic law determines what prerequisites are required for the domestic validity of a treaty. In monist States, where international law and domestic law are part of one legal order, treaties are domestically valid as soon as they are duly ratified.

these constitutional provisions it could have been argued that 'the substantive norms which form the *core* of the Convention may be assumed to be directly applicable' (Nys et al. 2007). Taking into account the context, the object and the purpose of the treaty – they are said to be unconditional and sufficiently precise to be applied as such in a particular case and to provide the basis for a specific decision (Andorno 2005).

Finally, as mentioned above, medical research and practice have been substantially affected by entrance to the EU in 2004. According to Article 91 paragraph 3 of the Constitution 'if an agreement, ratified by the Republic of Poland, establishing an international organization so provides, the laws established by it shall be applied directly and have precedence in the event of a conflict of laws.' This provision expressly establishes the direct applicability of the EU primary legislation, such as treaties and regulations. Although scientific research and public health do not fall into the EC's exclusive competence,[27] EU law plays a crucial role in the development of these areas, through secondary legislation and the entire *acquis communitaire.*[28]

In the course of the harmonization process, EC provisions relating to research have been introduced into statutes, in particular the directive 2001/83/EC on the Community Code relating to medicinal products for human use, the directive 2004/23/EC on certain technical requirements for the donation, procurement and testing of human tissues and cells and its implementing directives 2006/17/EC and 2006/86/EC, and last but not least the directive 98/79/EC on in vitro medical diagnostic devices. These specific provisions are complemented by the data protection directive 95/46/EC, which sets standards for the processing of personal data. None of these acts deals specifically with genetic data or biological material, but they aim at the standardisation of procedures, which would facilitate the free movement of

---

Examples of monist States include Belgium, France, Japan, the Netherlands, Portugal, the Russian Federation, Spain, Switzerland, most Latin American countries. In dualist States, international law and domestic law are parts of separate legal orders. Thus, treaties have to be incorporated into domestic law in order to become domestically valid. There are two distinct kinds of dualist States. On the one hand are Germany, Italy, the US, and several Central and Eastern European countries (e.g. Poland). On the other hand there are the UK, a considerable number of countries which used to be part of the British Commonwealth, and the Scandinavian countries. In the first group of dualist States, formal parliamentary approval is sufficient to incorporate a treaty into domestic law. A treaty is treated as international law even after its incorporation into domestic law with the result that the treaty can be applied directly within the domestic legal system. In the second group of dualist States, parliamentary approval is not formal but takes the form of substantive implementing legislation. The treaty loses its international law character in this process and, therefore, cannot be applied directly. See: Kaiser (2009).

[27] The competences of the EU are divided up into exclusive or shared competences, with other areas where the EU may take action only to support, coordinate or complement member states activities. Those areas that are not specifically mentioned in the Constitution remain the responsibility of the member states.

[28] The *acquis communautaire* is the entire body of European legislation, including all the treaties, regulations and directives adopted by the European Union (EU) and the rulings of the European Court of Justice that each new country joining the EU is required to accept.

goods and services. This means, medical professionals in Poland do not operate in a legal vacuum. Paradoxically, deregulation which is an important aspect of liberal governance seems to employ an abundance of regulatory instruments.

5.3 Genetic Research

5.3.1 Genetic Research as Research Involving Human Subjects

5.3.1.1 Setting the Scene

Like in any other developing field, most of the work surrounding medical genetics is still in its experimental phase. In times when genocentric dogma still prevails, genetic research has attracted a lot of funding. In July 2008 the world's largest cancer biobank was launched as a part of a population screening project conducted by the Cancer Hereditary Centre (Wysocka 2007). It is expected to contain approximately 2,000,000 personal records and an estimated 185,000 DNA samples (in Poland 20,000 participants), including diagnosed carriers of mutations for breast, ovarian, large intestine, prostate, and uterine cancer. Polish scientists have been very successful and have managed to patent new markers for genetic disorders causing breast cancer. The company administrating the DNA bank entered the Stock Exchange in July 2008. Still, in light of the absence of specific provisions regulating this area of research, they have been operating within the general legal framework established by the Polish Constitution (1997), primary and secondary legislation.

In comparison with other European states the Polish Constitution, as a relatively new document, provides advanced and comprehensive protection for individuals. Importantly, it contains explicit reference to medical experimentation. First of all, it guarantees the protection of human dignity (Article 30), freedom of scientific research (Article 73) and right to chose and pursue one's occupation (Article 65). Secondly, it prohibits medical experimentation without the individual's consent (Article 39). Finally, it provides procedural guarantees in case of rights' violation. Hence, a patient, doctor or researcher 'shall have the right to appeal to the Constitutional Tribunal for its judgment on conformity to the Constitution of a statute or another normative act upon which basis a court or organ of public administration has made a final decision' (Article 79). Furthermore, everyone has the right 'to apply to the Commissioner for Citizens' Rights for assistance in protection of his freedoms or rights infringed by organs of public authority' (Article 80).[29] Obviously, as an act forming the basis of the state's system the Constitution does not provide detailed provisions either on medical research, or medical practice. Therefore, a closer look at the primary and secondary legislation is needed.

As genetic research usually involves tests of biological material taken from individuals and subsequently analysis of personal data, unless samples are

[29]Constitution of the Republic of Poland, 2 April 1997, Dz. U. Nr. 78 poz. 483.

unidentifiable, general rules on research with human beings will apply. They are contained primarily in the Act on Medical Professions (1996),[30] The Pharmaceutical Law (2001),[31] The Act on Medical Devices (2004),[32] the Act on the procurement, storage and transplantation of human cells, tissues and organs (2005),[33] and the Personal Data Protection Act (1997).[34] The main bodies regulating the genetic research would be the Minister of Health issuing regulations and statutory instruments, medical research committees, the Agency for the Registration of Medicinal Products, Medical Devices and Germicidal Products, the Medical Devices Commission (advisory body),[35] and the General Inspector of Personal Data Protection.[36]

More detailed deontological norms regarding genetic research and practice are to be found in the Code of Medical Ethics adopted by the Polish Chamber of Physicians and Dentists in 2004.[37] They constitute one of the most important standards of professional conduct, the violation of which may lead to professional liability. Nevertheless, as the Polish Constitutional Tribunal stated in 1993, in the case of conflicting norms, while the doctor may behave in accordance with legal provisions, she/he might simultaneously breach the provisions of the Code. Such a breach cannot lead to disciplinary proceedings (Lenczowska-Soboń 2003).

5.3.1.2 Types of Medical Experiments with Human Subjects

The rules concerning medical experiments are regulated in Chapter 4 of the Act on the Medical Professions (1996).[38] According to Article 21 of this Act, medical experiments conducted on human beings can either be of a therapeutic or scientific nature. A 'therapeutic experiment' aims to introduce new or partially proven

[30] Act on the Professions of Physicians and Dentists 1996 (later also: Medical Professions Act or MPA) (Ustawa o zawodach lekarza i lekarza dentysty, Dz.U. 1997 Nr 28 poz. 152).

[31] The Pharmaceutical Law Act, 6 September 2001 (Ustawa prawo farmaceutyczne, Dz.U. 2008 Nr 45 poz. 271).

[32] Act on Medical Devices, 10 April 2004 (Ustawa o wyrobach medycznych, Dz.U. 2004 Nr 93 poz. 896).

[33] Act on the procurement, storage and transplantation of cells, tissues and organs 2005 (also: Cells and Tissues Act or historically: Transplantation Act 2005) (Ustawa o pobieraniu, przechowywaniu i przeszczepianiu komórek, tkanek i narządów, Dz. U. 2005 Nr 169, poz. 1411).

[34] Act on the Protection of Personal Data, 29 August 1997 (also: Personal Data Protection Act 1997) (Ustawa o ochronie danych osobowych, Dz. U. 1997 Nr 133 poz. 883 with later amendments).

[35] Both established by the Act on the Agency for the Registration of Medicinal Products, Medical Devices and Germicidal Products, 27 July 2001 (ustawa o Urzędzie Rejestracji Produktów Leczniczych, Wyrobów Medycznych i Produktów Biobójczych, Dz.U. Nr 126, poz. 1397 oraz z 2002 r. Nr 152, poz. 1263).

[36] Established by the Personal Data Protection Act.

[37] Polish Chamber of Physicians and Dentists, Code of Medical Ethics, 2 January 2004. Polish version to be retrieved from http://www.nil.org.pl/xml/nil/wladze/str_zl/zjazd7/kel.

[38] Articles 21 – 29 of the Medical Professions Act, *supra* note 32.

diagnostic, therapeutic or preventive methods in order to achieve direct benefit for an individual subject to the experiment. It can be carried out when previously applied methods are ineffective or their effectiveness is insufficient. A 'scientific experiment', aimed mainly at broadening medical knowledge, can be performed when participation in it is not bound with risk or this risk is inconsiderable. Both types of medical experiment can be performed when the anticipated benefits are considerable, and when the expediency and methods, are justified in the light of contemporary knowledge and are consistent with the principles of ethics (Article 22). A similar provision has been formulated in Article 27 of the Criminal Code (1997).[39]

Furthermore, the MPA (1996) bans scientific experiments on a 'conceived child' (Article 26 para. 3). This dubious category is understood to include foetuses and embryos, in vivo as well as in vitro. This interpretation fits in line with Article 45 para. 3 of the Code of Medical Ethics, which allows therapeutic experiments only in situations where expected benefits for the 'human being in embryonic stage of development' substantially outweigh the risks of not conducting it. These provisions obviously provide a higher standard of protection than the ones set in the Oviedo Convention (1997), which bans only the creation of embryos in vitro solely for research purposes. It seems also inconsistent with the provisions of Article 26 para. 1 & 2, which allow for scientific experiments on pregnant women, if they pose a minimal risk to the health of the woman and the conceived child. An interpretation reconciling the two principles suggests that Article 16 MPA (1996) should be read as to allow only research that does not involve the embryo. Still, it has been criticised for imposing higher standards of protection on embryo research than experiments on newborns (Boratyńska and Konieczniak 2001).

In this respect 'genetic research' is difficult to classify. It seems that experiments involving genetic material directed towards the discovery of new gene's function, population genetic screening and genetic tests will mostly fall into the second category of medical research, i.e. scientific experiments. In the case of population genetic screening especially, it would be difficult to prove a direct therapeutic effect on the individual participant. Still, even if such a link can be made, the evaluation of risk will meet substantive obstacles. The objective of a genetic experiment on humans could be: (a) the introduction of a new clinical method, (b) prevention and methods related to public health, (c) the examination of new medical devices and finally, (d) the examination of new diagnostic products.

Genetic research includes the so called pharmacogenetics and pharmacogenomics. The former is generally regarded as the study or clinical testing of genetic variation that gives rise to differing response to drugs, while the latter is the broader application of genomic technologies to new drug discovery and further characterization of older drugs (Evans and McLeod 2003). The issue of clinical trials for new therapeutic products is regulated by the Pharmaceutical Law (2001), which is

[39]The Criminal Code, 6 June 1997 (Kodeks karny Dz. U. Nr 88 poz. 553).

to be treated as *lex specialis*[40] to the Medical Professions Act (1996). According to Article 2 para. 2 of the Pharmaceutical Law (2001), a clinical trial is 'every experiment with human beings which aims at: (a) discovering or confirming clinical and pharmacological, effects of one or more therapeutic products or, (b) identifying unwanted effects thereof, or (c) tracking the absorption, metabolism or excretion of one or more therapeutic products in terms of their safety and efficiency'.[41] Therapeutic products are defined as 'substances or mix of substances in a pharmaceutical form of active substances or placebo' (Article 2 para. 2c). Although the term pharmacogenetic/pharmacogenomic has not been defined in the act, it is reasonable to believe that it means research on 'variations in DNA sequence as related to drug response'[42] and as such will fall into its scope.

According to the requirements defined in the Order of the Minister of Health on the Detailed Description for the Requirements of Good Clinical Practice (2005) the trial must be justified by earlier pre-trial findings, scientifically justified and described in the research protocol, based on ethical principles, conducted by a competent person guaranteeing a proper quality in a research centre.[43] Should the clinical examination of a therapeutic product constitute a life or health hazard to the participants of the examination, or should it be executed inconsistently with the approved protocol, then the Minister of Health will make a decision regarding its discontinuation. However, since pharmacogenetic research in Poland is still relatively limited, these regulations are not of major relevance.

The research and development of genetic diagnostic devices and test-kits will fall under the Act on Medical Devices (2004).[44] The scope of the Act includes medicinal products; (a) medical devices containing or derived from tissues or cells of human origins, (b) in vitro diagnostic medical devices manufactured and used within the same health care centre or used on premises in the immediate vicinity of this health care centre; (c) in vitro diagnostic medical devices intended to be used solely in scientific research. Outside the scope of the Act remain medical devices containing blood derivatives.[45] This type of research still falls within the scientific experiments, as they do not bring immediate therapeutic results for the research participant.

---

[40] *Lex specialis*, is a doctrine relating to the interpretation of laws in situations where norms collide. The doctrine states that a law governing a specific subject matter (*lex specialis*) overrides a law which only governs general matters (*lex generalis*). The situation ordinarily arises with regard to the construction of earlier-enacted specific legislation when more general legislation is later passed. This principle also applies to construction of a body of law or single piece of legislation that contains both specific and general provisions.

[41] The Pharmaceutical Law (2001).

[42] European Medicines Agency, Definitions for genomic biomarkers, pharmacogenomics, pharmacogenetics, genomic data and sample coding categories, November 2007, EMEA/CHMP/ICH/437986/2006. Retrieved on 30 August 2008: http://www.emea.europa.eu/pdfs/human/ich/43798606en.pdf.

[43] § 2 Order of the Minister of Health on the Detailed Description for the Requirements of Good Clinical Practice, 11 March 2005, Dz.U. 2005 Nr 57 poz. 500.

[44] Act on Medical Devices.

[45] Act on Medical Devices.

According to Article 29 Medical Professions Act (1996), medical experiments can only be executed following a positive opinion from an independent bioethics committee, the composition of which is determined by the Regional Chamber of Physicians and Dentists, the chancellor of a medical school and the chair of a research centre. The responsibility of issuing detailed regulations for the appointment and functioning procedures of such committees lies with the Minister of Health. A bioethics committee will not authorise research which contravenes the legal provisions set up in the Medical Professions Act (1996), the statutory instruments issued by the Minister of Health and the National Chamber of Physicians and Dentists.

5.3.1.3 Informed Consent

Article 39 of the Polish Constitution stipulates that 'no one shall be subjected to scientific experimentation, including medical experimentation, without his voluntary consent.' This provision needs to be interpreted in accordance with Article 30 and 31 that guarantee the protection of the inherent, inalienable and inviolable dignity of the person (Article 30), which constitutes the source of personal rights and individual freedom (Article 31). Although Article 39 mentions mere "consent", this principle has been developed further in the Act on Medical Professions (1996) and Pharmaceutical Law (2001).

According to Article 24 in conjunction with Article 25 para. 1 of the Medical Professions Act (1996) any medical experiment involving human subjects requires their written consent. Each research participant has to be fully informed by the doctor about the aims, methods and conditions of the research, expected therapeutic and scientific benefits, risk and their right to withdraw their participation at any stage of the experiment. If this is impossible, oral consent is given in the presence of two witnesses. The concept of informed consent is wider than that of the legal capacity of natural persons. Even individuals having limited legal capacity can still give valid consent as long as they are able to understand the nature and consequences of the experiment (Zielińska 2008: 410). In 1993 the Polish Constitutional Tribunal took the view that biomedical experiments that do not serve a therapeutic purpose and are undertaken without an explicit express consent are illegal.[46] Finally, research participants may withdraw their consent at any point during the project. However, if a disruption or withdrawal from the experiment may pose a danger to health, the doctor is obliged to inform the participant of this possibility (Article 24 para. 2 MPA).

Genetic research raises substantial difficulties with regard to minors. A child's involvement in such experiments may reveal either susceptibility to a disease with no clear indication of the risk of developing the actual illness, or a genetic disorder that will remain asymptomatic for many years, but which is at the same

[46] Polish Constitutional Tribunal, 17 March 1993, W 16/93, Dz.U. Nr 23, poz. 103.

time fatal. The Medical Professions Act 1996 requires that any medical intervention on a minor is conducted subject to prior informed consent granted by his/her legal representative. According to Article 98 of the Family Code[47] this refers to a parent or a person under whose custody the child is. The Medical Professions Act (1996) takes into consideration different phases of child's development. Therefore, it requires the doctor to obtain a 'cumulative consent' from both a patient who is 16 years old (or even younger if capable of giving consent) and his/hers legal representative (usually one or both parents). However, since a bigger involvement of the minor is foreseen in the decision making process, the statute contains detailed regulations on solving possible conflicts between patients, doctors and legal representatives (Article 25). In most cases these problems are solved by family courts.

Some commentators suggest that in light of the Family Code's provisions, which leave to the court the most important decisions regarding a minor's life, parent's consent to medical experiment is generally insufficient (Nestorowicz 2004: 142). In the context of research involving genetic diagnosis this suggestion becomes particularly interesting. On the one hand, with the exception of gene therapy, this kind of research will usually not be particularly dangerous for the child. On the other hand, growing uncertainties linked to genetics may be extremely distressful for the whole family. In the context of the paradigm change it becomes doubtful whether institutionalising such decisions would be beneficial to the child.

Consent should always be informed and specific, i.e. the aims, risks, benefits, intended outcomes and methods of the particular experiment should be explained. Therefore, no blanket or open consent is possible. It should be obtained in a written form or in the presence of two witnesses (Zielińska 2008: 411). Consequently, all new purposes that may occur in the course of the experiment need to be authorised by the research participant. Such a condition may be seen as restrictive, costly, inefficient and thus unsuitable for the purpose of genetic research which tends to be uncertain and unpredictable. However, in light of the recent paradigm shift and the criticism it induced, this solution may appear much more appreciated.

5.3.2 Research on Biological (Genetic) Material and Data

5.3.2.1 Rules Concerning the Use of Biological Material

As genetic research deals predominantly with the role of genes in the development of cancers, mental illnesses, and infertility, it involves biochemical analysis of biological material and subsequent study of the information derived from it. The initial row of letters forming genetic code may be translated into information about susceptibility to disease, gene penetrance, or genetic composition of a population. The fundamental regulatory difficulties with regulating the use of DNA concern

[47] Family Code, 25 February 1964 (Kodeks rodzinny i opiekuńczy, Dz.U. 1964 Nr 9 poz. 59 with later changes).

the anticipation of its dualism. The chemical substance and genetic information are inextricably linked and as such they escape the historic (Roman law) distinction drawn between rights over things and rights over persons. The problem occurs, because the law still seems to pursue this division. On the one hand it regulates the participation of individuals in medical research and the use of 'tangible' biological material (tissues, cells, organs), and on the other hand the use of information, knowledge and creations of the mind (data protection law, intellectual property rights). This is one of the reasons why genetic exceptionalists promote the idea, that use of DNA in genetic research requires special regulation. This problem was seen as particularly acute in the context of DNA biobanks, the proliferation of which was striking at the turn of the twentieth and twenty-first centuries.[48]

Although the problem of the legal status of genetic material and rules concerning biobanks attracted enormous attention in legal literature internationally (Bovenberg 2006, Kaye 2006, Knoppers 2005),[49] it has never been extensively discussed in Poland. Therefore, the use of biological material and data derived from it, fall under separate regimes. It seems to be assumed and uncontroversial that the mere fact of holding a biological sample cannot be tantamount to processing data. Data protection law can cover only information 'extracted' from samples and translated into scientific data. This approach is reflected in the way consent forms are formulated. The consent form presented to research subjects will include both consent to medical intervention, i.e. the collection of the biological material and consent to the collection and processing of personal data. Such consent will thus include authorisation of the violation of bodily and mental integrity as well as privacy of a person.

The use of biological material is therefore primarily regulated by the Polish Act on the procurement, preservation and transplantation of cells, tissues and organs (later: Cells and Tissues Act or CTA 2005)[50] the Medical Professions Act (1996), the Laboratory Diagnostics Act (2001)[51] and the general provisions of the Civil Code (1964).[52] They do not regulate the governance of *genetic* biobanks. However, the main difficulty may stem from the fact that they do not distinguish between biobanks created for research and other purposes. Prior to the amendments introduced in 2005 the Cells and Tissues Act applied only to organ transplantations. The provisions concerning the testing of biological samples collected in tissue and

---

[48] '[I]ts raw material is derived from people; its "product" is often the genetic information derived from analysis of this raw material; there are conflicts over the legal status of both raw materials and the product; and the product has no ready substitutes', in: Andrews (2005).

[49] See also: C.Grand/K Atia-Off, *Genmedizin und Datenschutz*, w: S.F. Winter/ H. Fenger/ H-L. Schreiber (red.), Genmedizin und Recht, Monachium 2001, pp. 538–540, Brückl (2001), or Halasz (2003).

[50] Act on the procurement, preservation and transplantation of cells, tissues and organs (Ustawa o pobieraniu, przechowywaniu i przeszczepianiu komórek, tkanek i narządów), 1 July 2005 (Dz.U. 2005 Nr 169 poz. 1411).

[51] Laboratory Diagnostics Act 2001, supra note 11.

[52] The Civil Code (Ustawa kodeks cywilny), 18 May 1964 (Dz.U 1964 Nr 16 poz. 93, last amendment Dz.U.07.82.557).

cell banks apply to 'investigations aimed at determining the usefulness of cells, tissues and organs for transplantation in human beings'.[53] Therefore, although the Act permits the procurement of cells, tissues and organs for research purposes, it seems that the establishment of a biobank solely for research purposes will not fall under the scope of the CTA 2005. Hence, it will not require a license from the Minister of Health subject to a positive opinion of the Cells and Tissues Authority and the National Banking Centre, but rather a permission granted by a local research bioethics committee under the Medical Professions Act 1996.

The Cells and Tissues Act 2005 concerns the procurement of biological material from living and deceased persons subject to prior specific and written informed consent, which can be withdrawn at any stage until transplantation. The research participant needs to be contacted and should authorise any changes in the use of his/her bodily material. The National Banking Centre should keep a record of every donor (and recipient) and the coding system employed for anonymisation purposes. These rules do not address the concerns raised by some commentators especially in relation to population genetic biobanks, i.e.: (1) participants' ongoing relationships with such projects; (2) control over access to data and biosamples; (3) participant and public influence over the use of such resources; and (4) altruism and trust safeguarded by independent ethical oversight, openness and accountability (Campbell 2007). The problem of benefit-sharing also remains unresolved and may take a particularly acute form in a highly individualized society that for historic reasons puts minimum trust in state actions.

However, while doubts are expressed as to whether genetic research can fulfill its promises, Polish law appears to provide a flexible and very liberal framework for the governance and management of research biobanks. On the other hand, as the alliance between science and power is complex, uncertainties surrounding genetics coupled with the breadth of information collected in biobanks may raise serious concerns about the accountability of scientists and industry involved in research as well as the democratic nature of experts' control provided by ethics committees (Plomer 2008). Therefore, even in the context of departure from genocentric dogma, the Polish legislator should pay more attention to the regulation of cell and tissue banks for research purposes.

5.3.2.2 Rules Concerning the Use of Biological/Health Data

Genetic material contained in cells, tissues and organs would be useless if it were impossible to translate the row of letters constituting genetic code into meaningful information. Data that allows the calculation of risks and probabilities linked to inheritance play a pivotal role in every kind of genetic research. Thus, the vast majority of legal analysis concerning medical genetics has concentrated on the issue of privacy and data protection. In research projects involving collection and analysis of DNA the two regimes (cells and data protection) apply subsequently.

---

[53] Art. 2 Cells and Tissues Act 2005.

The Polish Constitution guarantees everyone the right to the protection of private and family life (Article 47) and personal data (Article 51).[54] These two provisions allowed the Polish Constitutional Tribunal to develop the concept of the so-called 'informational autonomy', inspired by the German and Austrian legal theory. Therefore, in comparison to other European countries the protection of personal data seems to be wide and comprehensive. This protection is provided mainly by means of the Personal Data Protection Act (1997) (later also: PDPA 1997).[55] To a large extent the Act constitutes a literary translation of the EU Data Protection Directive 95/46/EC, which had to be implemented into Polish law.

The term personal data has been defined in Article 6 para 1 of the Act as 'data concerning an identified or identifiable natural person'. An identifiable person is one, whose 'identity can be determined directly or indirectly, in particular either through an identification number or through one or more circumstances that refer to his/her physical, physiological, mental, economic, cultural or social characteristics. The person is not considered identifiable, if obtaining this information would require excessive costs, time or actions.' The term excessive should be understood as 'unreasonable' or 'disproportional'.[56] This definition clearly embraces genetic data and suggests their processing is subject to control. According to Article 5, the PDPA 1997 constitutes *lex generali* in relation to other statutes, as long as they introduce higher protection of personal data.

At present, Poland has not finalised regulations or guidelines dealing with the use of genetic data. However, the PDPA 1997 explicitly distinguishes 'data concerning person's genetic code' as a separate category from 'data concerning individual's health'.[57] As both health and genetic data are classified as particularly sensitive, they are as a rule exempted from processing. Still, since Polish law fails to define genetic data it is not entirely clear what falls under this category, especially that 'health data' have been distinguished from 'data concerning the genetic code'. This distinction

---

[54] Art. 51 of the Polish Constitution:

1. No one may be obliged, except on the basis of statute, to disclose information concerning his person.
2. Public authorities shall not acquire, collect nor make accessible information on citizens other than that which is necessary in a democratic state ruled by law.
3. Everyone shall have a right of access to official documents and data collections concerning himself. Limitations upon such rights may be established by statute.
4. Everyone shall have the right to demand the correction or deletion of untrue or incomplete information, or information acquired by means contrary to statute.
5. Principles and procedures for collection of and access to information shall be specified by statute.

[55] Personal Data Protection Act 1996 (Ustawa o ochronie danych osobowych), 29 August 1996 (Dz.U. 1997 nr 133 poz. 883 z pózn. zm.).

[56] J. Barta, R. Markiewicz, *Ochrona danych osobowych. Komentarz*, Kraków: 2001, p. 505.

[57] Art. 27 para 1 Personal Data Protection Act, *supra note* 32.

would suggest that 'data concerning the genetic code' describe data derived from the non-coding parts of DNA, which usually serve identification purposes. These parts of DNA do not contain health information, and therefore cannot fall under the category of 'data concerning individual's health'. Such an interpretation, however, raises the question of the relationship between genetic data and other data such as eye and skin colour, race, and ethnicity, which have a genetic component. More importantly, it seems contrary to the *contextualist* view, according to which the sensitivity of the data depends highly – if not entirely – on the context in which they are processed (Simitis 1990). And yet, Polish provisions seem to be in line with the recent decision of the European Court of Human Rights in *S. and Marper v. UK*.[58] Although it concerns the use of data for forensic purposes it sets out informational privacy as a rule, the departure from which always requires strong justification.

Of course, the prohibition against the processing of sensitive data is not absolute. For the purpose of this paper, there are two most relevant exceptions from the general ban: (a) the consent of the data subject and (b) processing for research purposes. The way they are framed suggests they go beyond the standards constituted by the EU Directive 95/46/EC.

(i) Consent to Use Genetic Data

First of all, processing of sensitive data requires the written consent of the data subject.[59] The Polish Personal Data Protection Act (1997) defines consent as a declaration of will by which the data subject signifies his/her agreement to personal data relating to him/her being processed. According to Article 7.5 of the PDPA (1997) consent cannot be alleged or presumed on the basis of another declaration of will. Unlike the EU Directive, neither the term 'unambiguous' nor 'explicit' are present in Polish statute. Therefore Polish courts have been trying to specify the nature of consent. For example, the Supreme Administrative Court in its judgment stated: 'Consent for data processing is a declaration of will. This means that it should be analyzed from different points of view. Consent should be explicit. Lack of objection after providing data subject with information on intention of data processing does not constitute the explicit consent. (. . .) Data subject should know what kind of data will be processed, what kind of processing is intended and what the purpose of that is'.[60] In another verdict the administrative court of appeal stressed that consent should be 'precise'.[61] This means that the data subject should know the scope of consent. It is therefore necessary that the subject is provided with appropriate

[58] *S. and Marper v. U.K.* – 4 December 2008, Application no. 30562/04, [2008] ECHR 1581, available at: URL: http://www.bailii.org/eu/cases/ECHR/2008/1581.html. Retrieved on 26 April 2009.

[59] According to Art 27 para 2 consent is, however, not necessary for the erasure of data.

[60] 13.07.2006, I OSK 1083/05, LEX nr 275431.

[61] 15.11.2006, II SA/Wa 1612/06, LEX nr 301825.

information concerning data processing (i.e.: what kind of data will be processed, who will be responsible for processing, and what is the aim of processing).[62]

Generally speaking, consent cannot be alleged or presumed. Thus, it should be unambiguous and explicit. Consent that does not meet the above mentioned criteria is not valid. As mentioned above, the consent cannot be alleged or presumed on the basis of consent obtained in relation to a different matter. Therefore, following the distinction described above, the consent to participate in a medical experiment governed by the Medical Professions Act (1996) should be regarded as a different declaration of will than the consent given for data processing. It follows that consent given by the participant of an experiment does not automatically allow processing of the obtained data. As mentioned earlier, two consents are required, although they may be included in the same consent form. In the first declaration the patient consents to medical intervention. In the second declaration the patient consents to data processing.

The Personal Data Protection Act (1997) does not address the problem of consent given by minors. Article 27 para 2 states only that processing of the sensitive data is possible without consent when 'necessary to protect the vital interests of the data subject or another person where the data subject is physically or legally incapable of giving his/her consent until a legal representative is established.' Because there are no specific guidelines, general rules of the Medical Professions Act (1996) seem to be appropriate here. As Article 1 of the PDPA (1997) indicates that the Act applies to natural persons, it will not apply to deceased persons. The deceased person is afforded some level of protection through the Civil Code, which allows the person's family members to protect the person's good name against defamation. The rights rest with the person's family members.

This means that the problems with defining the data subject in relation to 'genetic data' that have occurred in relation to the Icelandic genetic data bank (*deCode Genetics*) remains unsolved in light of the Polish statute.[63] Article 25 PDPA (1997) stipulates the duties of the data administrator, in case personal data is collected from a 'person, other than the one they refer to'. In such a case, the data administrator is obliged to inform the 'data subject' of the purpose and scope of processing and the right to correct the data (although the wording of Article 32 does not anticipate whether the data administrator is obliged to disclose what kind of data has been collected). However, there are some exceptions to this rule. The duty is not imposed if the processing is allowed without consent or data is processed for research purposes (if the rights and freedoms are not violated) and informing the data subject would require disproportionate efforts. It is for the data controller to assess these circumstances.

The duty to obtain the data subject's consent constitutes only one out of nine exemptions to the general ban on the processing of sensitive data. Other legitimate

[62]Similarly administrative court in its verdicts given on 16.11.2005, II SA/Wa 139/05, LEX nr 213699 and 4.4.2003, II SA 2135/02, Wokanda 2004/6/30.

[63]For the discussion regarding the Icelandic biobank see: Gertz (2004) and Chadwick (1999).

purposes, such as research and public health, have been stipulated in Article 27 para. 2 PDPA (1997). This might suggest that processing for the purposes of medical research does not require written consent from the person, whose data is being processed. However, this interpretation will prove incorrect in cases of medical experiments, since in accordance with the Medical Professions Act (1996) (and Article 39 of the Polish Constitution) informed consent is required from all research participants. In cases where research relies upon genetic data, collected indirectly, by means other than from the research participant, data processing is possible as provided in PDPA (1997), even without informing the data subject if this would require disproportionate efforts. Such regulation in the context of genetic research may appear extremely liberal.

(ii) The Use of Genetic Data for Research Purposes

Although the Polish PDPA (1997) follows as a matter of principle the EU Directive 95/46 there are some differences in relation to scientific research. In the introduction to the Directive the states have anticipated circumstances where the use of sensitive data would be allowed 'in the substantial public interest'.[64] It may be worrying that neither the concept of public interest, nor the notion of scientific research has been defined. What is, however, more alarming is the way the Directive is implemented by national legislators and the state of data protection law in relation to genetic information in some EU countries, from which the Polish case may serve as the best example. As a general rule sensitive data should not be processed, except in certain situations, which would include research purposes. However, the exceptions foreseen in the Polish Act are much wider than those provided in the Directive, which makes the supposedly stronger protection of genetic data in medical research questionable.

In the light of Article 27 para 9 PDPA (1997) sensitive data can be processed (even without written consent) if the processing 'is necessary to conduct scientific research including preparations of a thesis required for graduating from university or receiving a degree; any results of scientific research shall not be published in a way which allows identifying data subjects'. It means that the research does not have to serve any 'substantial policy interest'. The sufficient criterion for the use of genetic data without consent of the data subject is the legality of processing. The PDPA (1997) will not be breached as long as the specific provisions of other statutes allow the processing of such data without consent and provide for adequate safeguards. However, since the notion of 'adequate/appropriate safeguards' has not been explained, this task will have to be decided by courts in particular cases, taking into account the proportionality principle. Due to an extensive set of rules contained in the Medical Professions Act (1996) the use of genetic data in the course of medical research, will usually acquire appropriate protection. Still, a bigger ambiguity seems to be related to the historic and non-medical (e.g. genealogical) research using

[64]Recital 34 of the Preamble to the EU Directive 95/46.

genetic data. It follows from the reading of Article 27 para 9 PDPA (1997) that in the case of publication of research results, the Polish legislator sees data coding as a sufficient mechanism for the protection of personal data. The appropriateness of such a solution has been questioned by scholars, who argue that privacy should not be understood as tantamount to confidentiality and anonymity (Chadwick 1999). This view has been that the data subject must be informed of the researcher's intention to use personal data.

According to Article 25 para 1 PDPA (1997) in cases where the data has not been obtained from the data subject, immediately after the collection of personal data, the controller is obliged to inform the data subject about the controller, the purpose and the scope of data collection, and in particular, details of the data recipients or categories of recipients, the source of data, the data subject's right to access and to rectify his/her data, and the right to object to data processing. However, according to Article 25 para 2 PDPA (1997) the data controller is not obliged to disclose the fact that data processing will occur as long as it is done for the purpose of scientific or historic research. This does not violate a person's right to be informed about the processing unless informing the data subject would require disproportionate efforts. Thus, it is for the data controller to assess these circumstances. The wording of Article 32 PDPA (1997) does not anticipate whether the data administrator is obliged to disclose the content of the collected data. Furthermore, unlike other national laws[65] the Polish act does not foresee any civil remedies for the breach of its provisions. All the foregoing problems may indicate that the concept of privacy, at least as realised in the existing norms, is endangered and does not give efficient protection to the subject of genetic information.

This vagueness is reinforced by the fact that, although according to the Medical Professions Act (1996) every piece of medical research requires authorisation by a local bioethics committee. Although Poland is a member of the Council of Europe, it has still not ratified the Convention on Biomedicine and Human Rights, nor any of the Additional Protocols. Thus, the only guidelines for Polish researchers in this respect are to be found in the Recommendation on the Protection of Medical Data (1997). However, due to the lack of a legally binding force, its impact is very limited. It still remains to be seen what legislative approach will be taken by the legislator in the future.

5.4 Medical Practice

5.4.1 Genetic Testing as Processing of Biological (Genetic) Data

As mentioned previously the use of genetic testing as a tool for diagnosing disease and predicting future disease risk has grown steadily. It is now used by some medical

[65]For example The Federal Data Protection Act, 18 May 2001 (*Deutsches Datenschutzgesetz*), published in the Bundesgesetzblatt I Nr. 23/2001.

professionals to establish which drugs would be most effective for an individual, based on his or her genetic variation. In Poland due to socio-economic and political changes highly professional state-funded clinical genetics centres are being challenged by a growing number of commercial labs, clinics and companies offering genetic testing with no further interpretation of results or counselling. The scarcity of legal regulations regarding genetic testing is matched by a reluctance of politicians and legislators to get involved in what is admittedly a difficult area. There is a growing interest in and demand from the public for genetic testing.[66] Genetic testing involves the analysis of genetic material and subsequent processing of data obtained as a result of this process.

The Personal Data Protection Act 1997 implementing the EU Data Protection Directive 95/46/EC, allows the processing of data concerning health and the genetic code when it is 'required for the purposes of preventive medicine, the provision of health care or treatment, where the data are processed by a health professional involved in treatment, other health care services, or the management of health care services, subject to appropriate safeguards.'[67] This provision constitutes an attempt to reconcile the privacy interests of patients with everyone's right to health protection and equal access to health care services, financed from public funds, irrespective of citizen's material situation outlined in Article 68 of the Polish Constitution.[68] The processing of health data by health professionals and civil servants managing the health care system is critical for its proper functioning. At the same time, it is difficult to determine what constitutes an 'appropriate level of protection' for the processing of health data. The formulation of the exception seems to suggest that confidentiality together with coding mechanisms and procedures are deemed to be sufficient. Consequently, patients' privacy may again be seen as tantamount to confidentiality and data security.

The possibility that data may be processed without individual consent also raises the question of whether such general rules would be sufficient to govern a prospective nation-wide health database consisting of medical and genetic data as well as banks of tissues, cells, and organs. As mentioned earlier, the management of biobanks has been specified by the Cells and Tissues Act 2005. Still, the Act refers to medical data only in the context of transplantation, enumerating the information that has to be collected at the time of tissue procurement and stored for 5 years.

---

[66]European Commission, CORDIS, EU project calls on Poland to pioneer new genetic testing guidelines, 21 September 2007, available at: http://cordis.europa.eu/fetch?CALLER=MSS_PL_NEWS_EN&ACTION=D&DOC=22&CAT=NEWS&QUERY=01204bc396f1:e508:045a3389&RCN=28383. Retrieved on 3 May 2009.

[67]It is argued that this section cannot be interpreted as basis for disclosure to insurers. Jackowski M., *Ochrona danych medycznych,* 2002, 116, J. Barta, R. Markiewicz, *Ochrona danych osobowych. Komentarz,* 2001, p. 435.

[68]In this respect Polish law seems to meet the criteria set by the Convention on Biomedicine and Human rights in Art. 3, which constitutes the right to equitable access to health care of appropriate quality depending on national resources. Convention for the Protection of Human Rights and Dignity of the Human Being with regard to the Application of Biology and Medicine: Convention on Human Rights and Biomedicine, Oviedo, 4.IV.1997, ETS 164.

Since organ transplantation and trafficking is ruled by a different rationale requiring the ability to be able to track donors' and recipients' cells, tissues and organs, no full anonymisation is desirable. Again, because the CTA 2005 foresees testing of samples only for transplantation purposes, its provisions do not apply to genetic testing conducted solely for diagnostic purposes (Siedlecka 2008). All kinds of testing procedures are governed by the Diagnostic Laboratories Act 2001. Hence, it applies also to genetic diagnosis. The Act centres on the principle of informed consent and professional qualifications and will be discussed below.

5.4.2 The Right to Know and the Right Not to Know

Polish law does not offer any specific rules concerning genetic information within the doctor-patient relationship. However, the Medical Professions Act 1996 (unlike British or German law) explicitly guarantees patients' right to know and the right not to know information about their health (Article 30 para. 1 and 3). The scope of the right to know encompasses information about past, present and future health, any genetic predisposition to disease as well as the diagnostic and therapeutic options open to the patient.

According to the guidelines provided in Article 51 para. h of the Code of Medical Ethics 2004, a doctor is allowed to conduct a genetic test only for health or research purposes after receiving the patient's informed consent and making genetic counselling available. However, since the latter contains deontological norms its breach may result only in professional liability. This means that as the law stands, health data obtained as a result of genetic testing may be used for other purposes such as employment and insurance, especially because Poland has not yet ratified the Convention on Human Rights and Biomedicine (1997).

The right not to know seems to play a subsidiary role, since it can be activated 'on the patient's demand'. Although the existence of such an explicit right should be viewed positively, the wording of the article does not solve the extensively discussed problem of how this right should be exercised in practice.[69] For in order to 'activate' the right the tested (to-be-tested) the individual has to acquire basic knowledge about the nature of the prognostic test, which may influence his choice, hence, limiting his/her autonomy. However, in this respect it is useful to invoke the German interpretation of the right not to know, according to which the information obtained prior to the patient's decision forms the basis of this decision and not the decision itself. In other words, one should not confuse the general knowledge about the nature of the genetic test – necessary to take the decision – with the information (knowledge) about the results of the test (about person's health). This approach complies with the wording of the Polish statute and seems to answer many doubts that have been raised in relation to practical relevance of the right not to know (Laurie 2002).

[69]For more discussion on the right not to know see: Stumper (1996), Andorno (2004) Taupitz (1998), and Damm (1999).

At the same time the Polish Medical Professions Act 1996 is silent when it comes to the correlative rights of patient's family members and doctor's obligations towards them. It stipulates, however, that the doctor can disclose any information (i.e. test results) to third parties only with patient's explicit consent (Article 30 para 2). This provision is followed by Article 40 para 1 which constitutes the confidentiality principle. It is interesting to note that according to Article 29 of the Code of Medical Ethics 2004 the doctor is expected to limit the disclosure of information about the patient's genetic code only to patients and their family members. Nevertheless, the Code does not provide any answer to the problem of conflicts inside the family. Moreover, neither the Medical Professions Act, nor the Code of Medical Ethics specifies the consequences of the doctor's breach of duty in this respect.

Finally, some doubts may emerge with regard to the possible conflicts between the Medical Professions Act 1996 and the Personal Data Protection Act 1997, which impose on the data administrator the duty to inform the data subject each time the data have been collected without their consent. Although, Article 5 of the PDPA 1997 contains a directive on the basis of which the rules of the Act may be treated as *lex generali* in relation to other provisions as long as the former guarantee a higher level of protection, it does not determine who should be the subject of this protection. This imprecision becomes even more ambiguous in the context of complex family situations, in which the notion of 'data subject' loses its semantic boundaries. Therefore, what seems to still be lacking is a complex, interdisciplinary and multi-dimensional professional training preparing health professionals for genetic counselling and follow up procedures. Outside the regulatory framework remains the whole area of 'genetic testing over the counter', which raises serious problems in terms of the protection of personal autonomy and privacy. This solution must be certainly welcomed by those who condemn genetic exceptionalism, determinism and reductionism.

5.4.3 Genetic Tests Available 'Over the Counter'

First of all, there is the difficulty of ensuring the *informed* consent when tests are offered directly to the public. This problem does not apply solely to genetic diagnostic tools. Although the Medical Professions Act (1996) and the Code of Medical Ethics (2004) provide such an opportunity, it is for the patient to decide whether he/she seeks advice or not. Despite these provisions, no institutional framework has been provided that enables patients to acquire such specialized interdisciplinary counselling. In times when genocentric dogma still prevails the absence of proper counselling poses the risk of adverse psychological, social and legal effects on the person and his/her family members. Companies involved in genetic testing may want to reinforce genetic determinism and mislead the public.

Since Poland has signed, but still not ratified, the Biomedicine Convention it is still not bound by the Additional Protocol to the Convention on Human Rights and Biomedicine, concerning Genetic Testing for Health Purposes, adopted by the Council of Ministers on 2 May 2008. It applies to genetic testing, rather than data, but its main focus is on the information derived from the tests, which can be regarded as biological (also genetic) data. Although the Protocol seems to reinforce to some extent the heavily criticised concept of 'genetic exceptionalism', it contains some guidelines concerning the way in which genetic testing should be done. Accordingly, it should be carried out only in response to a specific indication made on the basis of a precise evaluation, by a doctor. However, each state (signatory to the Protocol) has a large degree of discretion in deciding whether to allow a test to be carried out without individualised medical supervision. The Protocol prohibits exceptions to this rule in the case of tests with important implications for the health of: (a) the person concerned, (b) his or her family members or, (c) important implications concerning procreation choices. 'A precise evaluation of the situation of the person concerned, involving direct contact with him or her, appeared crucial in this regard: such an evaluation could not be carried out by means of a mere telephone conversation.'[70]

In Poland under current law this practice seems possible. The provisions of the Diagnostic Laboratories Act 2001 require that laboratories conducting DNA tests comply with quality, professional and legal standards. The diagnostic test must be undertaken with the patient's prior consent, unless diagnostic tests are prescribed by a doctor. The Act, however, does not require the labs to provide genetic or any other type of counselling. In this respect, the question arises of whether to make genetic tests available only under individualised medical supervision and whether this would enhance or limit the freedom of the individual autonomy.

Another question refers to the funding of genetic testing by the Polish National Health Fund (NFZ) and the role of genetics within the future health care system. Genetic tests are available, although they cannot be covered by the Polish NFZ. They are usually conducted as part of research projects and research screening programmes, like the one on hereditary ovarian and colon cancer. This may create serious inequalities in the future, if only wealthy patients can afford genetic testing, and concomitantly, more personalised preventive medicine. In Poland the lack of both funding and legislation stems from the financial and structural problems of the Polish health care system. Certain decisions in this respect would have to be made soon. However, although the existing Act on Health Services Funded by National Resources 2004[71] is currently being reviewed by Parliament, proposed amendments do not include genetic testing. It is unlikely that this will be changed by the end of

[70]Council of Europe, Additional Protocol to the Convention on Human Rights and Biomedicine, concerning Genetic Testing for Health Purposes, Strasbourg, 27 November 2008. Retrieved: http://conventions.coe.int/Treaty/EN/Treaties/Html/TestGen.htm.

[71]Act on Health Services Funded by National Resources, 27.08.2004 r. (ustawa o świadczeniach opieki zdrowotnej finansowanych ze środków publicznych), (Dz. U. Nr 210, poz. 2135). The Bill has been issued to the Parliament on the 5.06.2008 and is currently revised. See: Rzeczpospolita, Jakie świadczenia znajdziemy w koszyku, 5.06.2008, Retrieved on 08.11.2008, http://www.rp.pl/artykul/22,144184_Jakie_swiadczenia_znajdziemy_w_koszyku.html.

the legislative process. Yet, it still remains to be seen what, if any, steps will be taken to ensure the protection of individuals seeking genetic tests through intermediaries such as internet laboratories.

To sum up, Polish law does not provide any specific regulation of genetic testing or the use of genetic data in the medical context. In the Code of Medical Ethics 2004 only one provision has been devoted to the problem of genetic testing and counselling. Hence, overall there seems to be very limited concern about the use of medical data for medical purposes and the possible violation of individual autonomy and privacy. Although the Polish legislator expressly identified 'data about the genetic code' as sensitive data, the overall protection of the individual seems to be either comparable or even weaker than the protection offered by data protection acts in other countries, such as Germany, Switzerland, or the UK, where there is a proliferation of non-binding provisions issued by advisory bodies and institutions.[72] The biggest differences occur in the field of prenatal and pre-implantation genetic diagnosis, which generate the biggest controversies in the public debate.

5.4.4 Genetic Prenatal and Pre-implantation Diagnosis

Poland is best known for its very conservative and restrictive legal standpoint on abortion and the legal status of the human embryo. The abortion law embodied in the Act on Family Planning, Prenatal Life Protection and the Conditions under which Pregnancy Termination is Permitted 1993[73] has indeed been substantially influenced by Catholic doctrine and tradition. Interestingly, the fear of potential ostracism is so big that, many doctors are reluctant to perform abortion even in cases permitted by law. This is true even though termination is allowed only when: (a) pregnancy poses a serious risk to the life and health of the mother, (b) the pregnancy is a result of a rape, and finally (c) prenatal testing and other medical evidence indicates the high probability of a serious genetic disorder or incurable and lethal disease of the foetus. The ECHR judgment in the case of *Alicja Tysiąc*[74] demonstrated that the enforcement of the already restrictive abortion provisions is still very problematic in every day practice. Especially in small rural communities doctors experience a lot of pressure from the Church and Catholic organisations, which get directly involved in patients' decision making processes.[75] It may seem

[72]In Germany guidelines are issued by the Federal Associations of Physicians (*Bundesaertztekammer*), German of Human Genetics (*Deutsche Gesellschaft fuer Humangenetik*), and in the UK the Human Genetic Commission, Clinical Genetics Society, or Nuffield Council.

[73]Act on Family Planning, Prenatal Life Protection and the Conditions under which Pregnancy Termination is Permitted 1993 (Ustawa o Ustawa o planowaniu rodziny, ochronie płodu ludzkiego i warunkach dopuszczalności przerywania ciąży), 7.01.1993, (Dz.U. 1993 Nr 17 poz.78).

[74]ECHR, *Tysiąc v. Poland*, 20.03.2007 r., Application No 5410/03.

[75]Recently, a 14-year-old girl who had sought abortion was separated from her mother by a court decision (later revoked) and subjected to pressure from a priests and Catholic hospital manager. See: Bielesz (2008).

that this approach is also reflected in the field of genetic testing. This presumption is, however, only partially correct.

Similar to other countries, there is no exhaustive list of the diseases justifying abortion. Genetic counselling is required prior to and after the test is conducted. This is the only time where the duty to offer genetic counselling is explicitly imposed on health professionals. The case of *B.&S. Wojnarowscy*[76] in which a woman was denied prenatal genetic testing despite the fact that she already had given birth to two physically disabled children constituted another example of the discrepancies between the existing law and existing medical practice. The Polish Supreme Court in its landmark judgement on what constituted the first wrongful life case decided in 2005 that the couple were entitled to claim the costs of raising the third disabled child. Still, although these cases touch upon the issue of the ontological and legal status of the embryo in vivo and the autonomy of the woman, they do not deal with the problem of privacy, and the shared nature of medical data. Since the provisions of the Family Planning and Pregnancy Termination Act 1993 are silent in this respect, the general rules concerning confidentiality and the processing of information outlined in the Medical Professions Act 1996 apply.

This considerably restrictive approach in the area of abortion and prenatal genetic testing might raise the expectation that the status of the embryo in vitro and the preimplantation genetic diagnosis will be governed by similar rules. This is exactly what happened in Germany, where PGD is prohibited. In fact, because of the liberalisation of the German abortion law, at the moment embryos in vitro enjoy a much higher protection than embryos in vivo. In this respect Polish regulations appear to be much more coherent, as they provide a gradually increasing protection of subsequent stages of the development of human life.

Nevertheless, it is interesting to note that to date, due to the polarisation of the political scene, the government has been postponing the regulation of assisted reproduction technologies for years. The attempt to avoid the confrontation of Catholic and conservative values with the liberal tradition and practice has discouraged politicians from taking any legislative initiative in this respect until very recently (Jacorzynski and Kozlowski 2005). For the same reason it has not yet managed to implement the EU directives 2004/23/EC, 2006/17/EC and 2006/86/EC on procurement, testing and storage of human in vitro cells and tissues.[77] Therefore, due to the lack of any provisions concerning the procedure of IVF, the legal status of the embryo in vitro remains unregulated.

Hence, in light of legally binding norms, pre-implantation genetic diagnosis is permitted for all kinds of purposes, including sex selection and HLA tissue-typing. Since none of the IVF practices are regulated, private clinics would be able to conduct all kinds of tests. The provisions of the Medical Provisions Act prohibiting any

[76]The Supreme Court has decided on the admissibility of the claim in 2005. See: Sąd Najwyższy, 13.10.2005, IV CK 161/05.

[77]All available at: http://www.dh.gov.uk/en/Publichealth/Scientificdevelopmentgeneticsandbioethics/Tissue/Tissuegeneralinformation/DH_4136920. Retrieved on 3 May 2009.

experiments with the so-called 'conceived child' do not apply, since PGD cannot be regarded as an experimental method. The discussion currently taking place in the Polish literature focuses therefore on questions of whether IVF procedure can be considered a medical treatment and thus whether the state should contribute to it financially and finally, whether the right to procreation is an absolute human right or not (Haberko and Olszewski 2008). All these issues are extremely fascinating as they reflect the exact state of the legal debate concerning assisted reproduction techniques in Poland. However, they all exceed the scope of this paper.

The agreement reached at this stage is that in the absence of specific regulations, the provisions of the Civil Code and the Family Code will apply. What is more, Poland remains bound by supranational norms adopted by the EU. It is also doubtful if the above-mentioned provisions concerning genetic testing stipulated in the Code of Medical Ethics would apply. The fact that they require the patient's consent implies that they refer to postnatal genetic tests. They also lack legal force, which means their breach may result only in professional liability. The reason why Polish clinics are still not conducting PGD on a regular basis is that the procedure is still extremely expensive.

The obligation to implement the above-mentioned EU directives and the pressure to ratify the Convention of Biomedicine and Human Rights (signed in 1999) forced the government to undertake the legal initiative to introduce the necessary changes to the existing law. In spring 2008 the Prime Minister appointed a special ad hoc bioethics committee to prepare the ratification of the Oviedo (Biomedicine) Convention. While it was chaired by a conservative MP Jarosław Gowin, who is a devoted opponent of in vitro fertilisation, the committee aimed at regulating issues such as assisted reproduction techniques. However, unsurprisingly, due to fundamental differences among members, the committee was not able to reach a consensus. As a result the final report presented to the Prime Minister in October 2008 consisted of two separate parts, offering two opposing viewpoints on every subject discussed by the members. Since then the Prime Minister has not been able to make a decision. Interestingly, in the meantime the ruling party formed another committee – this time in the Parliament – to regulate assisted reproduction techniques. Alternative legislative initiatives at both sides of the political scenes have been undertaken. A civic initiative, supported by the Catholic Church, aiming at an absolute ban on IVF has managed to collect 150,000 signatures – much more than necessary for a bill to be discussed in Parliament. At this point it is impossible to predict the outcome of this initiative. Hopefully, it will finally give the Polish society the chance to experience an open, transparent and fully informed debate about the issues of genetics, biotechnology and bioethics.

5.5 Conclusions

The aim of this chapter was to demystify Polish law regulating medical genetics and present a coherent and comprehensive account of it in light of the changes towards post-genomics. The analysis showed that Polish provisions usually meet the standards set out in the international and European arena. Contrary to the predominant view, most of the Polish regulations are very advanced. The Polish Constitution provides a comprehensive bill of rights that includes expressly enumerated rights on the protection of personal data, private life, freedom of research, freedom of action, dignity, and the prohibition of medical experiments with individuals without their explicit consent. Moreover, the Act on Medical Professions 1996 elaborates on these general rules and details the provisions concerning medical research and practice. It is remarkable that it explicitly guarantees the confidentiality principle, the right to know, the right not to know, and the so-called 'therapeutic privilege'. All these rules comply with the provisions contained in the Convention of Biomedicine and Human Rights 1997. Therefore, its ratification will not necessarily constitute a revolutionary change of the legal state of affairs.

Furthermore, Polish law may appear better fitted to the shift towards the post-genomics era in life sciences. Polish regulators, being preoccupied with democratisation and liberalisation processes, have not introduced any specific regulations with regard to medical genetics. This means the Polish legal framework is 'anti-exceptionalist' and flexible. Such an approach may seem desirable in light of the latest developments in genetics and genomics. Still, what may be of concern is the fact that such a situation stems, not from an informed decision of the legislator, but rather from a lack of political interest in the subject of medical genetics and genomics in research and clinical practice. As a result the field remains unregulated and extremely liberal. The exceptions from the general ban on the processing of sensitive data stipulated in the Personal Data Protection Act 1997 go further than the personal data protection directive 45/96/EC, which may have adverse effects on the individual in the future. The problem of genetic testing 'over the counter' may take an acute form, not because of the possibility to determine someone's fate, but rather because of the confusion and misconceptions around genetic knowledge. Although genetics has already entered every day life, the general public does not seem to be aware of its implications, especially regarding the complex relationship between science, economics and politics.

The only sphere which draws public and political attention is genetic testing conducted on foetuses and embryos in vitro. However, it is not the human's enhancement that is at the centre of the controversies, but rather the moral and legal status of the pre-born human life. Consequently, very restrictive rules concerning abortion and pre-natal genetic diagnosis are contrasted with deregulation in the field of assisted reproduction techniques. Because of the genuine impossibility of reaching any kind of consensus between the few participants of the debate, the pre-implantation genetic diagnosis remains entirely unregulated. A normative vacuum in this field may create serious threats to personal autonomy and privacy at the vertical and horizontal level of a citizen's life as well as great benefits, of which the general public seems to be unaware.

However, it is not the lack of regulatory framework that raises the biggest concerns. Despite the delays in the implementation of several EU directives and the ratification of the Oviedo Convention 1997, advanced human rights provisions have indeed been incorporated into the main body of legislation, and in this respect

Poland should certainly not be seen as a country hindering European harmonization. These human rights could certainly become a platform for future cooperation. However, there is still a lot to be done in terms of democratisation of legislative processes and compliance with the existing human rights standards in every day practices. In order to achieve this goal, public awareness of the advances in life sciences and the surrounding discourses should be raised. Therefore, the most crucial and urgent need for Polish political elites is to pursue the creation of effective and transparent mechanisms of public debate. This would enable the public to make informed decisions in the increasingly complex and interdependent world of information, science and technology.

References

Ahn, A.C., M., Tewari, Ch.-S., Poon, and R.S., Phillips. 2006. The limits of reductionism in medicine: Could systems biology offer an alternative? *PLOS Medicine* 3(6):0709.

Alleyne R., and K. Devlin 2009. Genetic research in a "blind alley" in search for cures for common diseases. Daily Telegraph, 20 April 2009, Available at: http://www.telegraph.co.uk/health/healthnews/5189873/Genetic-research-in-a-blind-alley-in-search-for-cures-for-common-diseases.html. Accessed on 1 May 2009.

Andorno, R. 2004. The right not to know: An autonomy based approach. *Journal of Medical Ethics* 30(5):435–440.

Andorno, R. 2005. The Oviedo convention: A European framework at the intersection of human rights and health law. *Journal of International Biotechnology Law* 2:135–136.

Andrews, L.B. 2005. Harnessing the Benefits of Biobanks, *Journal of Law, Medicine and Ethics* 1(33):23.

Bielesz, M. and W. Lublinie wrze. 2008. 14-latka chce aborcji. *Gazeta Wyborcza*, 11.06.2008, available at: http://wyborcza.pl/1,91027,5300116,W_Lublinie_wrze__Czternastolatka_chce_aborcji.html. Accessed on 3 May 2009.

Boratyńska, M. and P., Konieczniak. 2001. *Prawa Pacjenta*, 117. Warsaw: Difin.

Bovenberg, J.A. 2006. *Property rights in blood, genes & data: Naturally yours?* Leiden: Martinus Nijhoff Publishers.

Brückl, D. 2001. *Rechtsfragen zur Verwendung von genetischen Informationen über den Menschen. Ein Beitrag zur Steuerung der Verwendung von Wissen*. Berlin: Wessensee Verlag.

Campbell, A.V. 2007. Ethical challenges of genetic databases. Safeguarding altruism and trust. *King's Law Journal* 18:227–245.

Chadwick, R. 1999. The Icelandic database-do modern times need modern sagas? *BMJ* 319:441–444.

Chadwick, R. 1999. Icelandic database. *British Medical Journal* 319:441–444.

Connor M. and M. Ferguson-Smith 1997. Essential medical genetics, 5th edn., 3. Chichester: Wiley-Blackwell. See also the Websites of: American College of Medical Genetics, available at: www.acmg.net or Cambridge University's Department of Medical Genetics, available at: http://www.cimr.cam.ac.uk/medgen. Accessed 1 May 2009.

Damm, R. 1999. Prädiktive Medizin und Patientenautonomie. Informationelle Persönlichkeitsrechte in der Gendiagnostik. *Medizinrecht* 10:439.

Evans, W.E. and H.L., McLeod 2003. Pharmacogenomics – drug disposition, drug targets, and side effects. *New English Journal of Medicine* 6:349–358.

Foucault, M. 1997. The birth of biopolitics. In *Ethics: Subjectivities and truths. Essential works of Foucault 1954–1984*, ed. P. Rabinow, 74. London: Penguin Books.

Gertz, R. 2004. Is it 'me' or 'we'? Genetic relations and the meaning of the 'personal data' under the data protection directive. *European Journal of Health Law* 1:231–244.

Gibbons et al. (1994). In: Rouvroy, A. 2008. *Human genes and neoliberal governance. A foucauldian critique*, 47. New York, NY: Routledge-Cavendish.
Griffiths W.M. et al. 2000. An introduction to genetic analysis, Chapter 1. In *Genetics and the organism*, eds. A.J.F. Griffiths, W.M. Gelbart, and J.H. Miller, et al. New York, NY: W.H. Freeman & Co. Available at: http://www.ncbi.nlm.nih.gov/books/bv.fcgi?rid=iga.section.60. Accessed on 28 April 2009.
Haberko, J. and K., Olszewski 2008. Moralne i prawne aspekty dążenia małżonków do posiadania dziecka w kontekście techniki zapłodnienia pozaustrojowego (in vitro). *Prawo i Medycyna* 30:65–80 and Jeszcze o moralnych i prawnych aspektach dopuszczalności zabiegów in vitro – polemika. *Parwo i Medycyna* 31. The response was from M. Boratyńska. 2008. Umowa o w sprawie zapłodnienia pozaustrojowego, *Prawo i Medycyna* 30:81–96.
Halasz, Ch. 2003. *Das Recht auf bio-materielle Selbstbestimmung*. Berlin-Heidelberg: Springer.
Harris, J. 2007. *Enhancing evolution*. Oxford: OUP.
Jacorzynski, W. and M., And Kozlowski 2005. Auf dem Weg nach (n)irgendwo: Die polnische Debatte zur Stammzellforschung. In *Crossing borders*, Grenzüberschreitungen, eds. W. Bender, Ch. Hauskeller, and A. Manzei, 469–487. Münster: Agenda Verlag.
Jassanoff, S. 2007. *Designs on nature. Science and democracy in Europe and the US*. Princeton, NJ: PUP.
Jenkins, H. et al. 2004. A proposed framework for the description of plant metabolomics experiments and their results. *Nature Biotechnology* 22:1601–1606. 6 December 2004.
Kaiser, K. 2009. Treaties, Direct Applicability, Max Planck Encyclopedia of Public International Law, OUP, retrieved on 27 April 2009 at: http://www.mpepil.com/subscriber_article?script=yes&id=/epil/entries/law-9780199231690-e1468&recno=1&searchType=Quick&query=dualist+state.
Kaye, J. 2006. Do we need a uniform regulatory system for biobanks across Europe? *European Journal of Human Genetics* 14:245–248. November 25.
Knoppers, B.M. 2005. Biobanking: International norms. *The Journal of Law, Medicine and Ethics. Regulation of Biobanks* 1(33):9.
Krajewska, A. 2009a. Conceptual quandaries about genetic data – a comparative perspective. *European Journal of Health Law* 16:7–26.
Krajewska, A. 2009b. Conceptual quandaries over genetic data – a comparative perspective. *European Journal of Health Law* 1:7–26.
Laurie, G. 2002. *Genetic privacy. A challenge to medco-legal norms*. Oxford: Oxford University Press. and Wertz, D.C. and J.C. Fletcher. 1991. Privacy and disclosure in medical genetics examined in an ethics of care. *Bioethics* 5:221.
Lenczowska-Soboń, K. 2003. Kodeks etyki Lekarskiej – moc prawna i stosowanie w orzecznictwie sądów prowszechnych. *Prawo i Zdrowie*. 13 July 2003, Accessed on May 23, 2008: http://www.abc.com.pl/problem/69/7
Maschke K.J. 2005. Navigating an ethical patchwork – human gene banks. *Nature Biotechnology* 23(5):539–545.
Nestorowicz, M. 2004. *Prawo medyczne*, 142. Toruń: TNOiK.
Nys, H., L., Stultiëns, P., Borry, T., Goffin, and K., Dierickx. 2007. Patient rights in EU Member States after the ratification of the convention on human rights and biomedicine. *Health Policy* 83(2–3):223.
Peltonen L., V.A. McKusick 2001 Dissecting human disease in the postgenomic era. *Science* 291, 5507:1224–1229, 16 February 2001.
Plomer, A. 2008. The European group on ethics: Law, politics and the limits of moral integration in Europe. *European Law Journal* 14(6):839–859.
Rouvroy, A. 2008. *Human genes and neoliberal governance. A foucauldian critique*, 21–27. New York, NY: Routledge-Cavendish.
Siedlecka, E. 2008. Będzie polska ustawa bioetyczna? *Gazeta Wyborcza*, 10.10.2008 and In vitro: Restrykcyjnie czy pragmatycznie? *Gazeta Wyborcza*, 28.10.2008. Accessed http://wyborcza.pl/1,75478,5859661,In_vitro__Restrykcyjnie_czy_pragmatyczne_.html

Simitis, S. 1990. *"Sensitive Daten" – Zur Geschichte und Wirkung einer Fiktion, Festschrift zum 65 Geburtstag von M.M.Pedrazzini,* 469–493. Bern: Stämpfli & Cie.

Somsen, H. 2008. Cloning trojan horses: Precautionary regulations of reproductive technologies. In *Regulating technologies. Legal futures, regulatory frames and technological fixes*, eds. R. Brownsword and K. Yeung, 222. Oxford: Hart Publishing.

Stotz, K.C., A., Bostanci, and P.E., Griffiths 2006. Tracking the shift to postgenomics. *Community Genetics* 9(3):190–196.

Stumper, K. 1996. *Informationelle Selbstbestimmung von DNA-Analysen*, 133. Frankfurt-Berlin: Verlag Peter Lang.

Taupitz, J. 1998. Das Recht auf Nichtwissen. In *Festschrift für Günther Wiese zum*, eds. P. Hanau, E. Loreny, and H.-C. Matthes, 583–602, Vol. 70. Berlin: Geburtstag.

UN General Assembly, Universal Declaration of Human Rights, 10 December 1948, 217 A (III). Available at: http://www.unhcr.org/refworld/docid/3ae6b3712c.html

Wysocka, M. 2007. Polacy w czołówce badań genetycznych. 20 *Puls Medycyny* 163, December 19, Available at: http://www.pulsmedycyny.com.pl/index/archiwum/8873/1.html. Accessed on 20 May 2008.

Zielińska, E. ed. 2008. *Komentarz do ustawy o zawodach lekarza i lekarza dentysty*, 410. Warsaw: Wolters Kluwer.

Chapter 6
Managing Trust and Risk in New Biotechnologies: The Case of Population Genome Project and Organ Transplantation in Latvia

Aivita Putnina

6.1 Introduction

The new biotechnologies – organ transplantation, new genetics, stem cell research – influence people's lives to a degree never known before. They contribute to a change in perceiving society, offering new ways to think about kinship (Strathern 1992), illness disease and body. Accommodating these technologies within society creates new risks, uncertainties and hopes that challenge existing notions of morality and justice (Godbout 2002). The aim of the paper is to explore the construction and management of risks and mechanisms through with the technologies become accepted and supported. I use the concepts of trust and risk/distrust as two poles between which attitudes, practices and discourses on technologies are performed in different arenas – everyday life, hospitals, scientific institutions and Parliament.

Trust is a much researched mechanism of reaching social security. Luhmann (1979) saw trust as a mechanism of social relationships able to reduce the social complexity inherent to modern societies. Both Parsons (1978) and later Luhmann (1979) describe trust as an affective feature characterizing familiar relationships. This way trust is always a particular and not a general relationship. Parsons believed that trust is based on shared values and goals and therefore thought that the doctor-patient trust relation is problematic. He claimed that patients need a base for trustful validation of doctor's competence rather than a capability of evaluating doctor's competence him/herself. However, later explorations of trust in sociology show that the issue is much more complicated.

Giddens (1991) universalises the concept of trust to "institutional reflexivity" which is given a central role in the period of late or high modernity. Giddens points at the reflexive and active nature of trust as opposite to blind and passive trust of "traditional" society. The traditional forms of trust are based in kinship, local community, religion and tradition. Modern institutions are connected to trust through abstract tokens and expert systems. This means that the attitude towards biomedical

A. Putnina (✉)
Centre for Bioethics and Biosafety, University of Latvia, Riga, Latvia
e-mail: aivita.putnina@lu.lv; putnina@policy.hu

P.T. Robbins, F. Huzair (eds.), *Exploring Central and Eastern Europe's Biotechnology Landscape*, The International Library of Ethics, Law and Technology 9,
DOI 10.1007/978-90-481-9784-2_6, © Springer Science+Business Media B.V. 2011

technologies is not a stable and given variable and depends on reflexive accounts of past experience of particular citizens and collectives. Giddens stresses the role of social solidarity between patients and doctors as a main engine for moving from traditional forms of trust to its more reflexive, negotiated and socialised form.

The modern society lives with the duality respecting and trusting systems at the same time experiencing a generalised scepticism towards institutions and abstractions. Risk and distrust become the components of high modernity. As Giddens (1991: 126) writes: "Thinking in terms of risk becomes more or less inevitable and most people will be conscious also of the risks of *refusing* to think in this way, even if they may choose to ignore those risks. In the charged reflexive settings of high modernity, living on 'automatic pilot' becomes more and more difficult to do, and it becomes less and less possible to protect any lifestyle, no matter how firmly pre-established, from the generalised risk climate."

I am interested in both dimensions; creation of trust and distrust in modern society in relation to biotechnologies. As the new biotechnologies challenge basic concepts of life and quality of life, the mechanisms of trust are impinged within the network of "old" and "new" forms of social relatedness and fragmentation. The technologies allow fragmenting bodies into organs, cells and molecules and exchange between these parts. As Strathern (1992) claims in relation to English kinship, it forces reconceptualising the concepts of persons and things, individual exchanges practicing the technologies becoming more personalized and persons accumulating parts belonging to other persons and thus creating new forms of relatedness. On the one hand, technologies generate particular relationships of affectedness and these relationships are sustained and activated through trust. This trust can be personal and institutional. However, it is still constructed around individual illness, uniting bodily condition, medical procedures and treatment relationships. On the other hand, as a public reality technologies generate particular trust/distrust attitudes of non-affected people. Here the mechanism of generating risk and trust is different as media and communication technologies serve as a bridge between the society and biomedical technologies.

In this paper I discuss the construction of relations created by notions of trust and distrust. This approach differs from that of Giddens or Luhmann as I use trust as an entry point seeing how trust and risk is utilized in imagining, living and altering relationships and knowledge, switching the perspective from institutions and agents to relationships activated by technologies and creating and involving various kinds of participants – persons, illness, disease, hospital units, genes, organs and science. Trust and risk play an important role to link and help see relationships and practices of technologies as a whole. Risk is an immanent part of this world since borders are constantly moved and "secure" systems of classification eroded and rebuilt on grounds not imaginable before the emergence of the new biotechnologies.

Genetic testing and organ transplantation not only use the already existing frames of knowledge and morality, mixing market and gift economy[1] but also create

[1] See. Titmuss (1970/1997) discussion on parallels of organ donation to gift and market economy.

new kinds of relations and contexts that differ from the traditional doctor-patient relationships and enactment of Parson's sick role. Genetic testing evokes illness as a collective property linking relatives. Genome projects mobilize the whole population. Donating DNA samples and medical history to a genome project creates the very "population" and makes the question of donation relevant to the public. Disease ceases to be a single, isolated entity located in one body. Change in perception of disease evokes not only a new understanding of illness but also imposes new duties and responsibilities and identities to the collectives created by genetic testing procedures.

Organ transplantation contrary to genome projects already offers new possibilities of treatment. It allows creating specific relations between a particular donor and a particular recipient but also makes a broader community generally but not particularly linked to organ donation via the presumed consent principle.

Both technologies are imagined as somewhat contradictory. They are still based on individual decisions and individual treatment. Meanwhile the technologies have become more public both in terms of treatment procedures and the social relations they produce. For example, cancer diagnosis and treatment can be an individuated act posing moral dilemmas to the patient. Seeing cancer as genetic illness mobilizes kinship networks[2] and opens new moral questions of dealing with illness that occur as probability but not a symptom. "Having an illness" or affectedness depends not only on medical procedures but also on the decisions of persons involved.

6.2 Positioning the Technologies

The most challenging implication of the technologies, perhaps, is the erosion of the borderline between science and society, social and biological. According to Wynne (2005) one can roughly trace two social science approaches to the challenges posed by new technologies. One of them sees society and science as separate fields and deals with knowledge transfer between the two. Reaching some kind of equilibrium between the two helps to achieve an "informed" public, thus sharing responsibilities of the effects of technologies.[3] One can also give equal analytical (at least) attention to local and scientific knowledge and looks at the conflict between the two. Wynne suggests also "a third way" of prehension combining both approaches.

The other approach follows an anthropological tradition looking at epistemologies of society and science (Latour 1993, 1999, Lock et al. 2000, 2002, Strathern 1988, 1991, 1992). One sees the science and society divide as a consequence of historical nature/culture separation and views the new technologies as a means

[2]Finkler (2000) discusses how genetics reinforces family links and defines kin contrasting it with the notion of kinship being conceptualised as love and choice. As Finkler says, faulty genes give kinship another dimension (ibid: 181).

[3]Tutton, 2007, for example, stresses the development of participatory techniques as response to growing anxiety over technologies.

of creating new networks or relations between the parts or actants. From this perspective not risks and uncertainty but different kinds of relations that are created do not fit within the "old" separate domains of science and society, nature and culture. As Latour (2003: 36) claims, the second or reflexive modernity propagated by Beck (1992) is characterised by increasing awareness that control over actions is impossible rather than with the increased level risks. Haraway (1997) pictures the same process through image of cyborgs showing the proliferation of the products of new technologies beyond the set borders of nature/culture.

I am interested in creation, management and accommodation of uncertainties and risks of both technologies in Latvian society. Organ transplantation is an already accommodated technology established in 1973 around kidney transplantation. Heart transplantation commenced in 2003 thanks to the enthusiasm of cardio-surgeons. A debate over liver transplantation has been started, as Latvia is one of the few EU countries that does not include it in the public health system. Latvia also has initiated the Human Genome Project in 2001. It is presented as future technology and does not have any implications on clinical procedures as yet. Nevertheless it is run in close association with clinics and addresses clinically researched mono- and multi-factorial diseases adding one more dimension to researching particular diseases.

Seeing science as an objective realm of biotechnology development, risks emerge in the field of application of science to society showing the vulnerability of hard science outside the laboratory. Risks of organ trade (Scheper-Hughes 2004) or partial diagnosis of genetically determined disease (Franklin 2003) or leaking data on possible donors (mentioned in the Latvian case study) are located within the social but ontologically risks arise within incommensurability of both domains. The management of risk thus is allocated in both realms. The "social" mechanism of risk management is trust and law. It is both local and interpersonal. The "biological" mechanism for managing trust is procedural. Organ transplantation depersonalises organs to make them "spare parts" and the genome project anonymises donors making genotypes and phenotypes.

Alongside trust and distrust, gift and market economy are the final paradigms of analysis I want to introduce. Mauss' (1967) theory of gift economy has been widely used in analysing the new biotechnologies.[4] One can see the link between transactions and the kind of social relations (re)created by those transactions. Generating trust is the result of reciprocity. Sahlins (1972) distinguishes several forms of reciprocity, gift being a form invested with a greater trust between the participants of transaction. It is also the most stable and reliable form of transaction. The more distant a social relation is, the more insecure and uncertain the transactions become. So, risk can be seen to arise from misbalance in all kinds of transactions and not only one inherent in modern institutions. In the case of biotechnologies reciprocity works on already established doctor-patient relationships subjected to treatment procedures

[4]See Frow (1997), Rabinow (1999), Franklin and Tutton (2001), Godbout (2002), Titmuss (1970/1997), Waldby (2002).

and informal rules of mutuality. Both technologies modify the transaction involving new actors – donors, organs, family and broader public. As Waldby (2002: 309) argues following Mauss (1967) and Frow (1997), circulation of biological gifts involve forms of social reciprocity and imagined community in much the same way as the circulation of other kinds of material goods in both traditional and market economies.

Titmuss (1970/1997) discusses the effects of tissue donation and management in the light of gift and market economy. He claims that altruistic donation establishes ties of indebtedness between citizens and creates the community between strangers. Selling, to the contrary, creates an instrumental, non-binding relationship between producers and consumers. There is no tie between the bodily fragment and the person from which it is derived – the bodily fragment becomes a commodity and is incorporated as an object of possession without creating a tie between the vendor and the purchaser.

I see trust and distrust as emerging from coping with challenges the technologies pose to the "traditional" system of classification and positioning and relating of new actants – organs or genes/DNA detachable from persons (Franklin 2003, Strathern 1991, 1992). Stability of the classification system provides ontological security or as Giddens (1991) would say – trust in abstract tokens. New biotechnologies erode previously stable categories and divisions between illness/health, individual/collective, body/machine, science/society, nature/culture diffusing risk and trust outside the science and health care system. Seeing biotechnologies through the process of reciprocity positioned between gift and market economies allows seeing its inherent ambivalence (organs and genes are seen as a gift but commodified through the process of transplantation and collection). Public participation and governance become important forms of transactions managing the technologies and aiming to bridge the gap between gift and market. Case analysis shows how trust and distrust are negotiated through using resources of morality and rationality and addressing challenges posed by the technologies.

6.3 Methods

This chapter is based on data collected within the EU 6th framework project "Challenges of biomedicine". The Latvian case study is based on media text analysis and interviews with informants – persons directly or indirectly affected by the new biotechnologies.

Printed media analysis[5] allows capturing publicly circulated construction of risks as well as the means of managing the risks. The Media serves as a mediator between

[5] Both articles in Latvian and Russian – two most popular languages in Latvia – are analysed taking 4 largest Latvia daily newspapers "Diena", "Neatkarīgā Rīta Avīze" and "Latvijas Avīze", "Rīgas Balss" and Russian daily newspaper "Telegraf". Additionally news articles and comments in two most popular Internet news portals – Delfi.lv and Apollo.lv were examined within the same time frame.

science and society and framings of technology become the ways to think and live with the technology. 69 original texts on genetics and 53 original texts on organ transplantation were selected including news, interviews and electronic comments of readers- published between 1999 and May 2006 in Latvian mass media as well as speeches held in the Parliament of Latvia. All texts were selected through electronic archives using keywords 'gene', 'genetics', 'geneticists', 'genome', 'organ transplantation', 'donors', 'recipients', 'transplant technologies', 'transplant specialist'. Texts that contained general information on discoveries in the field, theories, scandals etc. reprinted from the media abroad were discarded. Stem cell transplantation and tissue transplantation articles were also included in the scope of analysis as those were described and referred to as "organ transplantation" not discriminating between organs, cells and tissues.

The media text analysis is then compared with 16 interviews and observations made in hospital wards and scientific institutions. Clearance for the interviews was obtained from the hospital unit and individual consent was asked from each individual respondent after explaining the project.[6] All interviews were performed in hospital settings or scientific institutions. Interviews allow seeing how risks and trust is created and accommodated via interpersonal transactions. Five interviews were conducted with kidney transplant patients and three with organ transplantation specialists. Both doctors and patients were selected with the help of the National Transplant Unit. Patients were approached through contacts with their doctors. A similar approach was used in selecting participants affected by genetic technologies. Four scientists and doctors participating either in the genome project or genetic testing were also selected. Only one interview was conducted with a genome project participant who was contacted through her hospital doctor and asked for an interview to which she agreed. Additionally, three breast cancer patients were recruited through a patient NGO as representatives of one of the genome project target groups. Interviews took 1 h 15 min on average, and were recorded and transcribed. Only one interviewee refused a recorded interview but agreed to express the opinion that was summarized after the interview.

All data was coded and processed using the Atlas.ti programme. This paper deals with codes directly linked to trust and risk as framed in the texts.

6.4 Biotechnologies and Risks

Organ transplantation and the human genome project appear in the media as ambiguous technologies. Even the most positivist-oriented media accounts are accompanied with readers' electronic postings subjecting the shear optimism on

[6]Clear social science research standards are not yet set up in Latvia. This research balanced between rather strict research clearance requirements in medical science and no formal regulations in social sciences, aiming at the institutional and individual support of each of the participants of the research.

development of science and thus confirming Giddens' (1991) thesis on skepticism inherent in trust towards institutions in high modernity. Readers' postings show that each initially positive message can be reinterpreted and doubted. Behind the bright development of national genetic science one can always see self-interest of scientists to abuse the donated data.

My interest rather lies in framing the risks[7] and relationships imagined within these risk constructions as well as emerging mechanisms of generating trust and negotiating risks. There are several grounds for framing and negotiating risks created by biotechnologies. One framing of risk is created from the positivist position and sees threats in ignoring the biotechnology development. The other position sees risks within unjust application of the technology – organ trade and criminalization being most often voiced risks.

6.5 Risks Produced by Technology: Living Without It

As Rabinow (1996: 137) notes, new technologies are increasingly justified not by accumulation of knowledge but clinical applications of the technology. Risks are a central part of arguing the practical utility of the technology, linking laboratory or clinic to society and creating the need for society. Inability to introduce new biotechnologies is constructed both at a national and individual levels. On a general level risk links new genetics to the nation and sees the failure of introducing national genetics project as "losing" the Latvian genome fund:

> Of course, it is not enough to fence off from such desirers [foreign companies wishing to profit on genome research of Latvian population] by law and not doing anything ourselves hoping that we will keep our genetic fund clean and unspoiled. (Diena, June 21, 2002)

The Genome project thus creates not only a DNA depository but also a social entity – a population with borders thus creating the need to protect the "population genome". This way risk production and utilization within the ideology of nationalism is an essential part of biotechnology promotion. Though organ transplantation also is presented as a national project of saving lives it is still seen as a project primarily designed for individuals.

> The risk of dying as a result of not having a [heart transplant] operation is manifold (..) Heart transplantation gives hope to a patient who would otherwise die because even most modern methods of [non-invasive] treatment do not save the life. One should understand that the patient feels that he is slowly dying. (..) Due to these indications, heart transplantation risk seems low and one sees only a bright future ahead. (Rīgas Balss, June 20, 2002)

However, the individual life-saving effort is multiplied through the increasing need for transplant operations as a response to increasing risks for modern life that cause "overuse of organs" and demand their "pre-timely replacement". Thus social life

[7] I use frame as a concept allowing to identify the set of arguments around constructing biotechnologies according to people's experiences, similarly to Gamson and Modigliani (1989).

and society is seen as producing risks that would not be possible to imagine without the help of technologies. At the same time risks are inherent to social life and cannot be solved just using biotechnologies. As Beck (1992), Wynne (1996) and Giddens (1991) argue, these socially produced risks also create growing skepticism towards scientific developments, for example, seeing transplantation technologies as risk for organ trade.

6.6 Risks Applying Technology: Individual and Collective

Another linkage of the biological and the social is established when speaking of the risks that genetic information creates in society. Most media articles and readers' electronic postings on genetics mention such risks as misuse of genetic information by insurance companies, employers and even general practitioners selecting "better" patients. The construction of risks is symmetric. The benefits are seen in creating people's knowledge of potential illness as opposed to the risks if the information leaks in public space putting the question of power and control at the heart of risk production:

> Both sides are right. On the one hand, it would be very good to warn a person that he has a high risk of some illness and therefore should avoid everything that can promote that illness. Here is a simple example – your neighbor can risk smoking. But you cannot as you HAVE elevated level of risk of getting lung cancer. On the other hand, the probability that an employer use this data to his or her advantage is rather high in Europe and this is an important problem. (Krabis, readers' comments to delfi.lv, October 15, 2003)

What creates risk is the discrimination at the workplace or public sphere in general. Linking, discrimination at the workplace or insurance to the genome project makes the project risky. Linking it to health prediction at an individual level makes it beneficial as the responsibility over one's health has a positive value:

> The second aim [of the genome project] is establishing risk factors influencing "Latvian" genes. It is interesting, as already established, that environmental hazards do not affect everybody. For example, most smokers will never contract lung cancer. But there are some kinds of genes that are inherited and that guarantee almost 100% that lung cancer will develop. (Latvijas Avīze, February 5, 2003)

In fact there are many more social probabilities to interpret the situation. For example, an employer can perceive genetic risks in a beneficial way as guidelines for ensuring good health practices for his or her employees. It is the social context (or image of employer as a potential discriminator) that makes the release of individual data risky. This context drives the human genome project towards strengthening the measures against misuse of the data. Efforts are put in anonymizing and bundling data to create a collective genetic profile of the population. At the same time social responsibility of employers where the risk arise is ignored.

Interestingly, private initiatives – a business project carried out collecting umbilical cord stem cells featured in media or the National Geographic genome project on human ethnogenesis[8] mentioned in one interview – were not considered as carrying the same risks as the population genome project. Individual level allows seeing benefits of the technologies and making sense to participants' lives: the promise to utilize stem cells for future individual treatment or contribution to the story of world ethnogenesis that opened a new aspect for informant's interest in development of humanity. The difference in these particular projects is that they address particular social fears (responsibility for the child's health and wellbeing) or beliefs (the common human descent story) while the human genome project advocates vague and collective benefits.

The genome project creates new risks that are perceived as such in relation to organ transplantation where illegal organ trade is the most popular fear of the application of technology. As Rabinow (1999) has pointed out, biotechnologies create new alienable parts of the human body. Human bodies, boundaries, body fragments are changing from organs to cells, DNA, separable, re-exchangeable and reincorporable body parts (ibid: 95). Alienation is activated through images of losing these parts of the body. Through giving a renewable bodily part, blood donation can lead to further fragmentation of the body:

> Some rich men waiting for the donor organ can use my donated blood sample to perform all tissue matching procedures. All antigens, DNA – because you give blood with all the cells. After donation one risks of being run over by a car in a dark street and becoming a victim in an accident in which a person without internal organs ran under the car. (Readers' comments, delfi.lv, April 11, 2002)

As Frow (1997), Rabinow (1999), Franklin and Tutton (2001) have noted, positioning organ transplantation between gift economy and market makes it ambiguous and risky. On the one hand the ethical application of organ transplantation technology is seen within the gift economy as "pure" gifts. On the other hand, organs are alienated from bodies and made "commodities". So, necessity for anonymity of donors is both backed by argument for excluding blackmail of donor's relatives. At the same time imagining organ transplantation as a gift requires keeping personal the relationship between the donor or recipient via obligations of giving, receiving and paying back the gift. The market alienates the organs from the donor and turns them into commodities (which organs become through anonymisation procedures) which the recipient buys (contrary to state organized organ transplantation procedure) thus cutting the link between the donor and recipient.

Risks are located within the ambiguity of turning organs from gifts to commodities. As Lock (2001: 73) investigating the history of creating alienable body parts writes: "Moral disputes will inevitably be implicated in the manipulation of human biological materials no matter to what extent efforts are made to transform these materials into autonomous, reified entities". The procurement of cadaver

[8]One of informants claimed participation in National Geographic project exploring ethnogenesis of world population. He paid share in this project and sent his sample collected as oral cavity smear according to the instruction provided. The point of mentioning this illustration is positive voicing of genetic project.

organs and the statement of death prior to the removal of organs is problematized neither in Latvian media nor interviews with medical staff and patients seeing it as a process of explicit rules and procedures. This allows transferring agency and responsibility from doctors to the anonymous set of regulations. The regulations are issued by the Cabinet of Ministers and the Ministry of Health and describe the standard procedure. In practice, as one of the interviewed doctors recognises, there is always a risk of being accused of inappropriate treatment causing the death but this doubt is never extended to the very fact of the death statement. The debate on brain death and its link to organ transplantation had not reached media and both interviewed doctors and patients perceived death as a solid fact stated by the doctors.[9]

6.7 Using Risks in Reconceptualizing Illness

Biotechnologies allow seeing biologically determined properties that become risks when socially applied. Thus the construction of risk, for example, links women with a combination of gene mutations creating breast cancer. At the same time testing becomes the means for reducing the risks using techniques of surveillance and examination[10]:

> Scientists have discovered two genes protecting against breast and ovary cancer. If these genes have mutated (changed) they do not fulfill their functions. The statistics show that 80–90% of women carrying these gene changes run the risk developing breast or ovary cancer. Doing genetic testing and screening for people carrying these mutations, carefully observing and deeply examining, one can timely detect and prevent the development of malignant tumors. (Latvijas Avīze, November 8, 2003)

These risks have always been present but genetic testing makes them explicit. The concept of risk grounded by genetics is sustained through evoking more familiar kinship and family ties:

> I understand, I have read somewhere, that nowadays there is, here in oncology or somewhere else, a so called 'risk group'. Through this line, this gene line, I think it applies to breast cancer – if you have it in your family, then you receive special attention, of course, if you

[9]Only a few articles in press through 2000-June 2006 mention the distinction between brain death or death in general. However it is never problematical, e.g.: "In general, brain dead people who had lost any hope to survive become donors" (Neatkarīgā Rīta Avīze, 01.11.2002) or "There is such a definition as brain death. (..) Misdetection of brain death nowadays is impossible – reflexes, symptoms, bioenergetic activities of brain are examined in order to establish whether brain still functions." (Rīgas Balss, 20.06.2002)

[10]One can push Foucault's (1973) argument on medical power, knowledge and surveillance even further seeing genetic technologies as the means for deepening the reliance on medical technologies. Patients' movement recovering their subjectivity and strengthening their position within the field of medical knowledge (Armstrong 1982) is much threatened by genomics as it relocates the authoritative knowledge in interpretation of genes – individual properties that however can read only via specific medical knowledge.

> seek help. They [doctors] pay more attention if you go to the doctor and [say that] your grandma or mom, or some other female relative has had the disease. (Interview with breast cancer patient, aged 53)

However, it does not mean that the genetic formulation of illness is perceived as a risk at all as it proposes a new model of transmitting disease. The sick role described in Parsons (1951) is based on perceiving illness or disease as a single and autonomous entity detached from the person having it. So, the informant – donor of the genome project and a doctor herself – reflects on the sudden seizure of her coronary disease despite the anamnesis of her family:

> Informant: My dad died from untreated infarct, my mother died from ischemic insult, my sister died from heart dysfunction and brother died from ischemic insult. So, I have a horrible anamnesis! But when they were alive stents were not available. Coronagraphs, for checking blood vessels, might have existed but were not widely available.
>
> Interviewer: And your youngest relatives – when they did learn about it, did they go for check-ups?
>
> Informant: No. I have only one son. He does not go anywhere. He is 38 but does not go for any health checkups because he feels good.

The son of informant disassociates himself from the disease similarly as his mother does. Avoidance of disease ignoring even prophylaxis is a common problem of the Latvian population.[11] Genetic information therefore can be simply rejected as it contradicts the tradition of avoiding disease simply ignoring it until the very moment it threatens the normal life routine.

Organ transplantation provides new risks that, contrary to genetics, are located within the application of technology. Older kidney transplant patients remark that during the first decades of transplantation death rates were high. In practice that meant seeing fellow patients dying. At the same time these risks are seen as inherent to the technology but not to the skills of the doctors.

Several informants said the decision to have a transplant was difficult. This risk also plays an important part in accepting a donation from a relative. Organ rejection after transplantation is featured as the most dangerous risk both in interviews and media articles. In the case of live donation the biologic failure to accept an organ also means the failure to reciprocate, making the sacrifice and gift meaningless. A 43 year old kidney transplant patient Arturs, describes having a transplanted organ as living "with a ticking time bomb".

> In fact you go and do things, you have everything and then – an explosion. That is all and you cannot live anymore. If wars begins and you.. We [organ recipients] can go and blow ourselves up. We cannot live an autonomous life. There are thousands of patients. Maybe somebody is in a worse [condition that me]. No drugs – and the kidney stops and you need dialysis, no dialysis – everything stops, it's over. You don't exist anymore.

---

[11] See Putniņa, 2004: 47–51, Zobena, 2005: 37–38. Survey data show that only around one third of the Latvian population attend check-ups more or less regularly, they do it most often when they already have health problems.

This dependence becomes especially dangerous in the situations of social instability and Latvian patients still experience this with the fluctuations in the state-compensated drug system and the rise of patients' co-payments.

6.8 Trust in Abstract Tokens – Reputation of Science and Medicine

If risks associated with technologies are perceived as somehow having unmanageable consequences and side effects, trust is the result of confidence in particular familiar relationships and reputation. Confidence in science and the high reputation of scientists allows people to see risks as being positioned outside science. However, institutional guarantees – procedures and law – do not play the main role in establishing trust. As the organ transplantation case shows, personal relationships, reputation and mutual trust play the main role in successful organ procurement. The trust is based on traditional doctor-patient relationships and addresses trust problems at the level of personal interaction. Non-affected people stay outside of this trust mechanism. They relate themselves to technologies via media articles or TV soap operas.

The Eurobarometer survey on biotechnologies (Gaskell et al. 2006) shows Latvia in the third highest position of trust towards scientists after Malta and Cyprus. Around 95% of Latvian respondents express their confidence both in university and industry scientists, the average EU level being 90% and 82% accordingly.

Trust is not an explicit issue expressed in the general commentary around the genome project. Only two articles and two readers' postings to different articles deal explicitly with the notion of trust. Usually this notion is implicit and unproblematised picturing scientists as the representatives of objective scientific activity. Most articles on both technologies show that the dominating opinions are being expressed by scientists and doctors and deal with the description of work of particular scientists. The mechanisms of trust generation mentioned in media are directly linked to the reputation of scientists. For example, a readers' discussion on the establishment of a private genome research company for starting population genome project[12] in delfi.lv in July 18, 2001 is based on the argument that private enterprise is trustworthy since it is established by well-known university scientists (readers' comment, Nico, 18.07.2001 11:49). Similarly the commercialization of the project is defended in the political arena, scientists being active guarantors of the project while public is imagined as a passive subject of biotechnology research:

> Those [informed consent guarantees] were included in the initial text of [the Human Genome] Act project but later those were exempted simplifying the Act and delegating the regulation to the Cabinet of Ministers. If not for that simplification, the false acquisitions

[12]The initial ambition of Latvian Human Genome Project was the collection of population genome database. Due to the lack of resources and diminishing excitement around significance of population databanks worldwide, the project aims at collecting 100,000 samples of 2,270,000 of Latvian population (see more information on http://bmc.biomed.lu.lv/gene/KASirgenomadatubaze.htm).

> of the authors of the Act in calculated misinformation of people and cheating on informed consent would not have been made. One should be invested with an extremely negative imagination to think about elaborated means of fooling people. Even more, these [acquisitions] are turned against the honorable developers of the Act in Parliament and our most famous doctors and scientists who are the creators of genome database. (Neatkarīgā Rīta Avīze, August 1, 2001)

The discourse voicing confidence in scientists is a part of a discourse on risk of rejecting biotechnologies. This way scientists are directly associated with the eventual benefits brought by biotechnologies – improvement in quality of life, state competitiveness, comprehensive health care system and the like.

Technologies generate trust through hope and descriptions of either successful genetic research or transplant operations:

> I want to get to these [stem] cells. I have survived two cancer operations and I am only 44 and will become a grandmother in May. But I don't know where can I go [to get help] (Diena, January 21, 2006, readers' comments, SANDR AUZIŅA)

Transplant technology, like the new genetics, has a positive background associated with particular doctors and scientists working in the field. This way building trust on a general level is seen as a two-sided process involving not only the technical procedure of donation but also morality and reputation based on the altruism of doctors and scientists and the trust and moral obligation of donors and their relatives.

6.9 Between Gift and Market

A salient question to consider is why organ and gene donation does not become normative practice. As Godbout (2002: 88) writes "conceiving the morality from its starting point of the gift stops one from falling into moralist conception of morality, external to the subject". What makes the gift non-normative is its spontaneity and established linked between the persons. Godbout reaches the conclusion that a gift does not set moral judgement criteria that are useful for an ethicist.

Imagining gene and organ procurement as gift giving in normative institutional settings provokes ambiguity since donation lies between two opposite social principles. Altruistic donation in a gift economy establishes ties of indebtedness between citizens and creates the community between strangers. As Strathern (1988) notes, power and gender are implicated in the exchange of objects as they are conceptualised as part of persons. Therefore a donated organ cannot become completely detached from the person of whom it used to be a part. This is true both for donor's relatives and recipients. Donation is the relationship that separates donors and recipients. Selling in a market economy creates instrumental, non-binding relationships between producers and consumers, completely detaching an organ or a gene from a person. As some authors have pointed out, organ transplantation as gift economy is never pure (Frow 1997, Rabinow 1999, Franklin and Tutton 2001) because of illicit and official buying and selling of organs. Latvian "horror" stories like "then healthy person X has been murdered and their heart removed" (*Readers' comments, Delfi, 19.06.2005*) are based on the logic of market exchange as well as personal and shared experience of "wild capitalism[13]":

> [The] supply and demand in the market is rising. Hence human organ and tissue transplantation is coming forward as an explicitly big problem at the moment.[..] the scale of human organ and tissue trade is speeding up. Criminal business is taking over this sector (Rīgas Balss, 28.03.2003).

All interviewed doctors noted that organ trade media coverage contributes to refusal rates. It is higher when Russian TV channels show films on the organ trade.[14] Franklin and Tutton (2001: 8) write that with the growing value of body parts, trust in the medical profession has been eroded as commercialisation diminishes trust. Organ trade attaches market value to organs and the presence of this value influences the judgements of donation.

Organ trade stories offer an easier way to frame organ donation. A gift creates a much more complicated relationship that involves mutual obligations. The organ trade establishes a different kind of mutuality whereby moral and spontaneous qualities of exchange are replaced by money. Latvian doctors' interviews provide accounts of relatives of potential donors offering the organs of their deceased family member for sale on an almost daily basis. I also witnessed one such phone call from a potential live donor during an interview.

Commercialization is linked to social inequality (Schepher-Hughes 2004). A survey conducted in Latvia indicated that Latgale region inhabitants – one of the poorest in EU – towards had positive attitudes toward legalization of organ sale with 41.6 % respondents supporting it (Latvijas Avīze, 21.05.2001).

Looking from the positions of power/knowledge (Foucault 1973, 1977) the situation of trust arises from a symmetric and balanced exchange where both parties participate in decision-making based on gift. Mauss (1967) conceptualises gift relationships as a set of procedures and obligations establishing a moral bond between the persons exchanging gifts.[15] Indeed, the donation often does not end only with the agreement for donation. Some relatives of the donor, according to the doctors' interviews, want to learn about the general characteristics of the organ recipients and some recipients of the organ assume certain moral obligations (like the obligation to

[13]Informants often use this phrase to describe the sudden comodification of life and health in particular after moving to market economy. During the Soviet regime health care was free and comparatively well financed by the state.

[14]I traced several scandals of tissue transplant possibly illegal procurement and export in Latvian press. The media does not differentiate between the tissues and the organs. Therefore even tissue procurement is a task of completely different authorities and regulations. The organ transplantation unit has suffered the consequences of the tissue scandal.

[15]Mauss distinguishes the gift giving (showing generosity and deserving respect), gift reception (showing respect to the donor) and the obligation to return the gift (showing the respect and honour the giver). Though Mauss have dealt with "primitive" societies were gift giving plays an important role, it has been both criticised (Testart 1998, Laidlaw (2000) and successfully applied to the Western societies (e.g., Polanyi 1957)).

live healthy life and make the best use of the donated organ). The return of the gift is also extended to society in general taking up the work for the public good. So, a kidney transplant patient, a woman aged 61, recalls her engagement in the patient NGO following the operation:

> So, had not I fallen ill, had not my kidneys not refused to work, I – of course, my life would have been different – but I think there is much more benefit for me now. [Smiles]. (..) Thanks to the operation I can help others now.

Trust should always be re-established as it is not only the condition but also the product of particular donations.

The organization of transplantation strengthens trust relationships. The trust generated by the transparency of the procedure was shared by all interviewed organ recipients and was based both on the belief that no-one can influence the complicated matching procedure and personal qualities of the procurement brigade staff. The trust of the participants is sustained by their face-to-face community. As the transplant waiting period is comparatively short (a maximum of a year and a half) and patients often know each other from haemodialysis procedures, trust relationships are managed inside the community.

It is possible to reconcile both the market and gift approaches, as suggested by one of the procurement brigade surgeons. He suggested that the state could provide compensation for donor funerals, where relatives would accept such help. The support, in non-monetary form, would conform to the obligation to "return the gift" by paying homage to the deceased donor.

> There have been cases in world practice when decisions have been made to cover the funeral expenses for donors but not give money directly. This is all linked to state budgets but this consists in fact of [only] a few hundred Lats for a family that has supported everything as compensation for funeral costs. Well, say, 50 donors per year and even if half of them would qualify for state support that would make the state poor. But then [human rights] protectors would argue that this is a human trade and market. I don't know; it is hard to say. We support the idea that people should receive compensation. Not as monetary compensation but to cover some funeral costs. (Organ transplant surgeon)

However, even if the quoted doctor's proposal comes from the logic of gift giving, it has to be translated into a monetary form at the level of state policy that uses the logic of market exchange. Looking from this logic such a proposition would violate the official "non-profit" character of the donation. However, the case study clearly demonstrates that particular exchanges are built upon the gift exchange principle and involve much more complicated mechanisms of building trust than transparency and disinterest of the sides stated by the law. Relation of donation is also expressed in arguing cost-effectiveness of transplantation opposed to artificial kidney. In fact two domains are fused here – market principle of exchange value and human life. As those are based on different moral principles they are not commensurable.

6.10 Conclusions

Risk and trust mechanisms provide the insight mechanisms that allow an examination of how the new biotechnologies are accommodated in society. The paper addressed the failure to accommodate the new biotechnologies within the existing social relationships of reciprocity (gift and market) and clearly divided borders of the social and natural (society and science). Risk emerges as misbalance in relationships and categories. Using the old traditions of establishing trust relations between the doctors and patients becomes more and more problematic as biotechnologies erode the traditional borders of clinical practice. Potential organ and gene donors are found outside the circle of directly affected persons. The notion of affectedness itself is extended outside the borders of the clinic and disease and becomes a matter of choice for participation in genetic screening or acceptation of organ transplantation as a cure from previous disability. People with renal transplants are officially considered disabled persons in Latvia while in Germany they are assumed to be cured. At the same time many kidney recipients in Latvia believed they were healthy. Transplantation science and genomics use socially produced images of disability and affectedness that do not describe the new in-between-situations of being potentially affected as in case of probable genetic disease or in case of probable organ rejection.

The construction of risk shows that risk is produced in social arenas outside hospital wards and scientific laboratories. Risks for organ trade or data leakage to the insurers or employers are risks generated in the social. Organ trade makes use of alienable body parts and discrimination takes genetic profile into account. Most elaborate data protection and tissue and organ redistribution measures within the fields of biotechnology application therefore will not diminish the risks created with conflating gift and market economies in the application of the biotechnologies. There is constant tension between the alienation of body parts and comodification through which the objectification of these body parts are reached. However, as Lock (2001) argues these parts have never been permanently detachable from the persons giving them.

Ignoring the broader links of technology to socially generated risks will always threaten the application of technologies. Reciprocity is a necessary component of social transactions. Thus not anonymity but the creation of social ties and a community of donors and scientists would allow for the establishment of trust. This analysis shows deep trust relationships among affected people in case of organ transplantation. The community of recipients is created through the time spent in hospital, interactions with medical staff and the very transplantation act creating "brothers and sisters in kidney" when recipients share the same pair of donor's kidneys. The organ transplantation case shows how important social stability is for organ recipients. They are mingled not only in the reciprocal relations with the donors but also in a wider network of health care policy and drug supply. As one of the interviewed doctors recalled, during early 1990s when the health care system was not functioning properly Latvian Prime-minister personally brought medicines for renal transplantation patients from his state visit to Taiwan. During this crisis doctors used their political connections to ensure the safety of their patients.

The traditional relations of trust also have their limits. Rather narrow circles of trust allow for the creation of trust networks in medical institutions where trust is based on already elaborate traditions of doctor-patient relationships. This tradition is based on passive patient and an active doctor position. Reliance on doctor's judgement and authority thus gives support to the technologies. However, biotechnologies are not seen as taken for granted choices outside the clinical setting when they compete with alternative explanations and perceptions of the body, illness and treatment. There are no reciprocal networks among gene and organ donors. None of the patients with a diagnosis of genetic disease explored within the genome project perceived his or her illness in terms of genetics. They saw the participation in the genome project as a favour to their doctor, hospital or as a way to understand their disease. Unlike transplantation, genetics does not provide the common perceived grounds for illness. Gene donors did not link the genetic cause to their particular illness. They remained loyal to their doctors but not to the genome project on the whole.

This aspect provides the basis for understanding the impact of technology on broader society. Trust relationships based on illness and application of the technology is sustained in face-to –face community, which makes it difficult to involve publics who are unaffected by illness or particular genetic conditions. There is no comprehensive system for establishing trust apart from legal framework and ethics as the means for ensuring the effective working of technology. Contrary to hospital settings where expert domination is justified through the treatment procedures and doctor-patient trust, these relations do not extend to the broader public – potential donors and their relatives. Law and the dominating ethical approach make potential donors unproblematic passive followers to the obligation to support science. With the trust credit to science being rather high one can ignore risk-talk. However, this abstract trust credit is constantly influenced by scandals and growing images of risk produced in the social realm.

Risks cannot be managed by ignoring the broader links of biotechnologies to society. Reciprocity is a necessary precondition of generating trust. Therefore not anonymity procedures but creation of reciprocal donor and scientist communities will create trust. Technologies unite too many different principles of morality to fall within one instrumental prescription of what is seen as morally correct. Therefore moral solutions can only be seen as the result of negotiations. Patient groups should play an important role in re-balancing trust and negotiating the social changes made possible by the introduction of biotechnologies. None of the newspaper articles examined here contained stories on transplant patient campaigns for organ procurement or patient group supported gene donation calls. Patient organisations still have great potential to shape the biotechnology debate in Latvia. Institutional regulation interferes and tries to make these moral relations abstract, losing the morality attached to particular relationships. New technologies can successfully operate only under the condition that a larger community, and not only ill people, feel affected or connected to biotechnologies.

References

Armstrong, D. 1982. The doctor patient-patient relationship: 1930–80. In *The problem of medical knowledge: Examining the social construction of medicine*, eds. P. Wright and A. Treacher, 102–122. Edinburgh: University of Edinburgh Press.

Beck, U. 1992. *Risk society: Towards a new modernity*. London: Sage.

Eliot, J., T., Foster, C., Hood, and L.R. Reed. 1997. Gossip cake and wine: Methodological challenges for feminist and dialogic research in cross-cultural settings. In *Invitation to dialogue: Beyond gender (In)equality*, ed. I. Koroleva, 11–33. Riga: Institute of Philosophy and Sociology, Academy of Sciences of Latvia.

Finkler, K. 2000. *Experiencing the new genetics: Family and kinship on the medical frontier*. Philadelphia, PA: University of Pennsylvania Press.

Foucault, M. 1973. *The birth of the clinic: An archeology of medical perception*. London: Tavistock Publications.

Foucault, M. 1977. *Discipline and punish*. Harmondsworth: Penguin.

Franklin, S. 2003. Re-thinking nature–culture: Anthropology and the new genetics. *Anthropological Theory* 3(1):65–85.

Franklin, S., and R. Tutton. 2001. Revisiting Concepts of Gift in the New Genetics, Report based on an international symposium funded by the Welcome Trust.

Frow, J. 1997. Gift and commodity. In *Time and commodity culture: Essays in cultural theory and postmodernity*, 109–217. Oxford: Clarendon Press.

Gamson, W.A., and A. Modigliani. 1989. Media discourse and public opinion on nuclear power: A constructionist approach. *American Journal of Sociology* 95(1):1–37.

Gaskell, G., et al. 2006. Europeans and biotechnology in 2005: Patterns and trends. A report to the European Commission's Directorate-General for research. Eurobarometer 64.3. http://www.ec.europa.eu/research/press/2006/pdf/pr1906_eb_64_3_final_report-may2006_en.pdf

Giddens, A. 1991. *Modernity and self-identity. Self and society in the late modern age*. Cambridge: Polity Press.

Godbout, J.T. 2002. Is homo donator a homo moralis? *Diogenes* 49(195):86–93.

Haraway, D. 1997. *Modest Witness@Second Millennium: The FemaleMan© Meets OncoMouse™*. London: Routledge.

Laidlaw, J. 2000. A free gift makes no friends. *Journal of the Royal Anthropological Institute* 6:617–634.

Latour, B. 1993. *We have never been modern*. Cambridge, MA: Harvard University Press.

Latour, B. 1999. *Pandora's hope: Essays on the reality of science studies*. Cambridge, MA: Harvard University Press.

Latour, B. 2003. Is *Re*-modernization occurring –and if so, how to prove it? A commentary on Ulrich Beck. *Theory, Culture and Society* 20(2):35–48.

Lock, M. 2001. The Alienation of Body Tissue and the Biopolitics of Immortalized Cell Lines. *In Body & Society* 7(2–3):63–91.

Lock, M. 2002. *Twice dead: Organ transplants and the reinvention of death*. Berkeley, CA: University of California Press.

Lock, M., Young, A. and Cambrosio, A. eds. 2000. *Living and working with the new medical technologies: Intersections of inquiry*. Cambridge: Cambridge University Press.

Luhmann, N. 1979. *Trust and Power*. New York: John Wiley.

Mauss, M. 1967. *The gift: Forms and functions of exchange in archaic societies*. New York, NY: Norton.

Parsons, T. 1951. *The social system*. Glencoe, IL: Free Press.

Parsons, T. 1978. *Action Theory and the Human Condition*. New York: Free Press.

Polanyi, K. [1957] 2001. The economy as instituted process. In *The sociology of economic life*, eds. M. Granovetter and R. Swedberg, 31–50. Boulder, CO: Westview Press.

Putniņa, A. ed. 2004. *Reproductive health of the population: Study on the situation in Latvia 1997–2003*. Riga: United Nations Population Fund (UNFPA) and Papardes Zieds.

Rabinow, P. 1996. *Essays on the anthropology of reason*. Princeton, NJ: Princeton University Press.

Rabinow, P. 1999. *French DNA. Trouble in purgatory*. Chicago, IL & London: Chicago University Press.

Sahlins, M. 1972. *Stone Age Economics*. Chicago: Aldine-Atherton.

Scheper-Hughes, N. 2004. Parts unknown: Undercover ethnography of the organs-trafficking underworld. *Ethnography* 5(1):29–73.

Strathern, M. 1988. *Gender of gift*. Berkeley, CA: University of California Press.

Strathern, M. 1991. *Partial connections*. Lanham, MD: Rowman and Littlefield.

Strathern, M. 1992. *After nature: English kinship in the late twentieth century*. Cambridge: Cambridge University Press.

Testart, A. 1998. Uncertainties of the 'obligation to reciprocate': A critique of Mauss'. In *Marcel Mauss: A centenary tribute*, eds. W. James and N.J. Allen, 97–110. New York, NY: Berghahn Books.

Titmuss, R. [1970] 1997. *The gift relationship: From blood to social policy,* eds. A. Oakley and J. Aston, 57–318. The New Press.

Tutton, R. 2007. Constructing Participation in Genetic Databases: Citizenship, Governance and Ambivalence in UK Biobank. *In Science, Technology and Human Values* 32(2):172–195.

Waldby, C. 2002. Stem cells, tissue cultures and the production of biovalue. *Health: An Interdisciplinary Journal for the Social Study of Health, Illness and Medicine* 6(3):305–323.

Wynne, B. 1996. May the sheep safely graze? A reflexive view of the expert-lay knowledge divide. In *Risk, environment and modernity: Towards a new ecology*, eds. S. Lash, B. Szerszinski, and B. Wynne, 44–83. London: Sage.

Wynne, B. 2005. Reflexing Complexity: Post-genomic Knowledge and Reductionist Returns in Public Science. *In Society Theory, Culture & Society* 22(5):67–94.

Zobena, A. ed. 2004–2005. Latvia. Human Development Report: Human Capability in Regions, Ed. UNDP, Latvia, ASPRI, 2006. http://www.lu.lv/materiali/petnieciba/resursi/human-development.pdf

Chapter 7
Social Trenches in the GM Food Battlefield: Experiences of a Survey Series in Hungary

Gyula Kasza and Zoltán Lakner

7.1 Introduction

Acceptance of genetically modified food has never been straightforward in Europe (Bánáti et al., 2007, Curtis et al., 2004, Gaskell et al., 2000), and many feared that EU accession of Central and Eastern European countries in 2004 would offer multinational biotechnology companies opportunities to introduce genetically modified products into communities by using their influence on the legislative systems of these states. Hungary was expected to be an exception: a country with a strong commitment to genetic modification (GM)-free agriculture. The traditionally strict food laws and the thorough inspection systems (hosting many parallel inspection processes done by five different and independent national authorities) developed under the old socialist system promised to guarantee the traceability and close control of products containing genetically modified organisms (GMOs).

Agriculture and food production provides: 5–6% of Hungarian GDP, employment for nearly 9% of the active population (data may be biased by the expectedly high, 30%, share of the shadow/informal economy, based on smallholder farmers and other agricultural activities that are not officially enumerated or taxed), 6% of total exports, and a basis for rural sustainability (Lakner and Hajdu 2004). Of course, the most directly affected part of agriculture in the GMO debate is seed production. With shrinking export opportunities, Hungarian seed production, based on traditional GM-free seed production, was expected to play an important role in agricultural export. The most important markets for seeds were the EU-member states, mainly Germany, Italy and Austria. At the beginning of the 1990s Hungary's share in EU seed import had been 6–8%, its value more than 150 million ECU. With this market share Hungary was the third most important exporter into the EU after the US and Canada; the next most important being Australia (3%) (Izsáki 2004).

The aim of this chapter is to analyze the social discourse on GM technology in Hungary. There is a wide range of knowledge on these debates in post-industrial,

G. Kasza (✉)
Corvinus University, Budapest, Hungary
e-mail: gyula.kasza@uni-corvinus.hu

P.T. Robbins, F. Huzair (eds.), *Exploring Central and Eastern Europe's Biotechnology Landscape*, The International Library of Ethics, Law and Technology 9,
DOI 10.1007/978-90-481-9784-2_7,

developed states (Santaniello et al., 2001) as well as in developing countries (Belt and Keulartz 2007). While there is a lively and often heated debate surrounding this issue in Hungary, its unique characteristics receive little attention and are almost unknown in the international science and technology studies community.

The chapter is divided into two main parts: the first offers an overview of GMO debates and the applicability of different consensus-seeking techniques; the second part analyzes the case of GMO regulation in Hungary. After the presentation of Hungary's official position, declaring it GMO-free, the views of different stakeholders are highlighted based on media analysis, individual in-depth interviews and consumer surveys. In the third part of the chapter, an evaluation and analysis of the Hungarian system is provided to explain the appearance, declarations and actual strategies of stakeholders in the GMO debate. The conclusion highlights the most important obstacles to public debate in Hungary on the GM issue. Such debate would be essential to forge a widely accepted and stable GMO regulation policy. Without it there is the danger that GM-related policy will become unstable.

7.2 Genetically Modified Agricultural Products as a Public Issue

The security of food supplies has and will be one of the most important questions in the history of humankind (Malthus 1820, Meadows et al., 1972, OECD-FAO 2008), and so solutions that offer safer and more stable agricultural production, even under conditions of climate change, can often be a focus of public attention. According to its advocates, genetic modification promises to solve global problems of malnutrition.

At the same time, there are considerable counter-arguments against GMOs. The controversy associated with genetically modified foods goes back more than 20 years (Nisbet 2006). The debate surrounding genetic engineering and its practical applications is highly complex (Janasoff 2005), and this complexity has been attributed to the number and variety of stakeholders, which potentially include all citizens with an interest in food (Priest and Gillespie 2000), energy production (Lemaux 2008), environment benefits and risks (Nap et al., 2003) and the problems of international trade (Paarlberg 2003).

Applying the general theory of factors affecting the acceptability of risks (the Psychometric Paradigm by Slovic (1987)), suggests that the public are less willing to accept perceived risks associated with consuming GM food, especially first generation GMOs, but in practice if hard evidence does not emerge demonstrating the reduction in consumption of GM foods the acceptance of these products will increase, especially if the disposable income of the population decreases (see Table 7.1). The social acceptance of the application of genetic engineering in agriculture should be examined in the context of risk perception and attitudes, as well as in public trust in regulatory institutions (Frewer et al., 2004).

There is a wide range of opponents to this new technology, accusing biotech companies of pushing and hastening the introduction of biotechnological innovations

Table 7.1 General theory of factors affecting acceptability of risks and the application of this theory to products of genetic engineering (GMO-specific factors are marked with: *)

Factors that make risks more acceptable	Factors that make risks less acceptable
Easily seen, visible	Not visible or easily seen (the consumer must believe the label)*
Known to the person exposed	Unknown to the person exposed (high degree of uncertainty in traceability)*
Immediate effect	Effect delayed (no short term ecological or health effect found, but debates focused on the long term)*
Old or familiar	New and unfamiliar*
Known to science	Unknown to science (people perceive it as too new to be known)*
Individual affected feels he/she is able to control the activity	Individual affected is not in control*
No feeling of "dread"	"Dread" (in peer-reviewed, grey and tabloid literature one can find a wide range of debate about the potential relation between cancer and consumption of genetically engineered products)*
Localised effect	Potential for widespread or global effect (concerns about an ecological catastrophe)*
Equitable	Not equitable (rich persons, or the citizens of rich countries can afford to buy organic or "GMO-free" products)*
Person who benefits bears the risk	Person who bears the risk is different to the one who benefits (especially in relation to 1st generation GM crops)*
Voluntary, self/exposed	Involuntary, imposed by someone else*
Individual	Catastrophic, has potential to affect many people at the same time*
	Potential for affecting many people at the same time
Low future risk	Perception that the genetic modification may cause high future risk (e.g. supposition of cancerous illnesses in generations to come)
Easily reduced	Not easily reduced with processing technology
Does not affect me	Affects me*

Source: Slovic, Fischhoff, and Lichenstein (1985) modified by the authors

into agriculture without analyzing the health and environment risks adequately. In this discussion both sides believe that ignorance is a key problem, and develop different strategies to "educate the public" – of course in line with their own arguments (Marris 2001). The different surveys highlight the low level of knowledge that consumers have on genetics (Gaskell et al., 2003). However, the increase of genetic knowledge does not necessarily diminish scepticism and polarization of opinions since people with different attitudes choose different arguments from a wide range of choice given to them by different sources of information from the academic sphere (Gaskell et al., 1998). Although public confidence in science seems to be altered profoundly, there are possible means to integrate the social sciences (and sometimes even the public itself) in future-shaping conversations and decision-making processes in the field of science and technology (Macnaghten et al., 2005, Nowotny et al., 2001, Wynne 2006, 2007).

GMOs became an important social issue before policy makers realized that public concerns needed to be addressed. Subsequently, public understanding of science became the new paradigm, which – in words, at least – differed from the elitist deficit model (Gross 1994), although they arose from the very same roots. Quantitative and qualitative studies revealed that efforts to foster scientific literacy, including in genetic engineering, by and large did not fulfil the motto that "to know science is to love it" (Janasoff 2005). Therefore, those assumptions that explained public unease with factors such as general ignorance, irrationality of lay people, NGO exaggeration and faulty or sensationalist reporting by the media, became contested. Wynne (2007) argue that although all of these exist, they do not explain public resistance when it occurs. Results of a comparative analysis of British and Spanish press on risk perceptions and attitudes towards GM food revealed that "no conclusive evidence can establish a direct cause-effect but only an association relationship between negatively biased news and the lack of public trust in the field" (Vilella-Villa and Costa-Front 2008: 2104.)

In analysing the GMO-related public debates in Germany, Switzerland, the United Kingdom and the EU, Birner and Alcaraz (2004) observe that the dialogues fostered transparency and understanding of the other party's opinion, still – as they remark – "no information could be found on whether or not participants changed their opinion during the process. Since the approach chosen allowed comparatively little interaction between participants, it is not designed to facilitate 'social learning'" (Birner and Alcaraz 2004: 21).

7.3 GM as a Public Debate

The importance of public participation in activities that have environmental consequences is widely acknowledged (Beck 1992, de Marchi and Ravetz 2001). There is an axiom that there is a direct, mutual relationship between the intensity of public participation and the development of civil society (Bryner 2001). Building on the work of Irvin and Stansbury (2004), we constructed a table of advantages and disadvantages of public participation in governmental decision-making. In the case of the GMO debate the situation is even more complex than is described in Irvin and Stansbury's (2004) original model, therefore we modified it even further (Table 7.2).

The public debates on different policy issues can be compared and evaluated on the basis of socio-cultural and political background at the level of the community concerned (national, regional, etc...) (McCombs et al. 1997). In the opinion of Villa, Spanish "lay people's ignorance with regard to science remains a historical legacy of 40 years of the authoritarian regime, during which the media were subject to censure. This may have led to an important cohort of the Spanish population being inhibited from expressing their views on strategic issues of Spanish agriculture such as GM food. Spain's 'new' democracy" (Vilella-Villa and Costa-Front, 2008: 2104).

Table 7.2 Advantages and disadvantages of citizen participation in governmental decision making (adapted from Stansbury (2004), modified by authors)

	To citizen participants	To government	Specific problems in the GMO debate
Benefits of citizen participation			
Decision process	• Education (learn from and inform government) • Persuade and enlighten government • Gain skills for activist citizenship	• Education (learn from and inform citizens) • Persuade citizens; build trust and reduce anxiety or hostility • Build strategic alliances • Gain legitimacy of decisions	• The most important sources of "scientific" information are research institutes, often financed by multinational biotech firms • The arguments of anti-GMO groups may increase the negative attitudes of average, formerly neutral citizens • Public attitudes towards government in general influence the acceptance of governmental arguments
Outcomes	• Break gridlock achieve outcome • Gain some control over policy processes	• Break gridlock; achieve outcomes • Avoid litigation costs • Better policy and implementation decisions	• The discussion freezes, and in place of real arguments about GMOs, unrelated issues of general dissatisfaction arise, such as the war in Iraq, that are connected by citizens with the GMO debate and political parties exert substantial influence in shaping debate outcomes
Risks/disadvantages of citizen participation			
Decision process	• Time consuming (even dull) • Pointless if arguments are ignored • Misrepresentation • Loss of legitimacy to oppose unwanted decisions	• Time consuming • Costly; • May backfire, creating more hostility towards government	• The potential negative long term effects of GMOs are rather hard to detect, and there will be (if any) damage to human health, environment or production only in the long run, opposed to the direct benefits (e.g. cheaper products) which – at least in theory – will be felt almost immediately
Outcomes	• Risk of decision being heavily influenced by opposing interest groups	• Loss of decision making control; – Possibility of bad decisions that are politically impossible to ignore	• The international pressure on national governments, intense lobbying by multinational firms can drastically change the final government policy

Spanish and Hungarian democratization processes show numerous similarities (Pickvance 1999), that is why we suppose that the lack of positive experiences of public debates on policy issues inhibits bottom-up initiatives in both countries. An important difference between the Spanish and Hungarian transitions is that during the last 2 decades in Hungary the basic foundations of a private ownership-based market economy had to develop too (Kornai 1992). Lack of a coherent transition strategy coupled with often chaotic and unsystematic privatization made the situation even more difficult (Spicer et al., 2000), further undermining civic trust in public institutions (Wedel 2003) and decreasing social capital (Angelusz and Tardos 2001). Given the hindered development of real "civil society", the different political parties try to manage each and every part of social life. For this reason, the Hungarian political system can be called a "partocracy" (Ágh 2001). In summary it can be stated that there are numerous hindrance factors of civil participation in public issues. The most important of these are historic traditions and the socio-economic framework. The "vacuum" in the social sphere is filled by political parties, which is why each and every social issue (e.g. the GMO problem) becomes subject to party politics.

Since 2000, GM food has become a popular media topic in Hungary. According to contemporary surveys (Bánáti and Kasza 2003), over 80% of Hungarian consumers are familiar with the notion of "genetically modified food", and reported mainly negative associations, although the meaning of the phrase was not entirely clear to people in most cases. Further opinion polls conducted in advance of the EU accession in 2004, also provided evidence for strong public concerns and resistance towards GM crops and food (Bánáti and Kasza 2003, Eurobarometer 2006, Lakner and Kasza 2005). Non governmental initiatives, primarily Ökotárs Alapítvány ("Eco-partner Foundation"), Élőlánc ("Living Chain") and Greenpeace Hungary also articulated strong criticism towards agricultural utilization of the known GM species.

Political parties did not address new biotechnologies during electoral campaigns in 2002 and 2006. Apart from a few television appearances, leading characters of the main parties did not express definite opinions on the subject. In rare cases, party experts employed in the background declared in informal "background" meetings with journalists, that "GM food and crops are to be kept away for as long as legally possible".

During years of hesitation and conflict-avoiding strategies by different governmental organizations and political parties, there was a broad opportunity for multinational biotech companies to apply different "infiltration techniques" with the purpose of introducing genetically modified products into Hungary. These techniques have been based on three pillars: (1) Strong integration into academic institutions through funding and scholarships. In 1999, the "Zoltan Barabás" Biotechnology Association was established with the participation of seven large-scale biotech companies (Monsanto, Pioneer etc.), numerous academic research institutions and – as private members – ten scholars of biotechnology. The aim of the Association is: "providing valid information to the public on biotechnology" (http://zoldbiotech.uw.hu/). (2) Successful lobbying at certain levels of governmental decision making and state apparatus, embracing practically all political parties and the opinion leaders. (3) Training and motivation of farmers. These agricultural producers, having above average farm sizes and qualifications, are important, influential and well-motivated promoters of modern biotechnology, accusing agricultural regulation as being "luddite". If they were successful in introducing GM varieties in Hungary and other new member states in the EU, their bargaining positions would be much more favourable with competent authorities of the Community. Although the EU gave its consent for the production of the GM corn variety MON 810, many European countries (Austria, France, Greece, Hungary and, effectively, Poland) have imposed bans. By the middle of 2008, even Romania, which had one of the most receptive markets to GM crops started to move toward a reversal of its stance. Attila Korodi, Romanian minister of environment said "we have to analyze the true costs of growing GMOs" (International Herald Tribune 2008). We may conclude from this that the GM critics' fear that CEE would become a "Trojan horse" that allowed biotech crops to be introduced to Europe through the back door has not become a reality.

7.4 The Hungarian Moratorium

All the factors mentioned in the introduction; the increasing market-side problems of Hungarian agriculture (from the time of the system change, the most important problem is not the increasing of production, but the selling of products) were expected to result in a mainly conservative, protectionist and risk-avoiding strategy regarding agricultural utilization of genetically modified crops. Accordingly, and joining a number of other European countries, Hungary announced a moratorium at a national level on the commercialization of GM crops from 20th January 2005. On 31st January 2006 this decision had to be re-approved by the Environmental Committee of the Parliament, on the basis of new studies.

After a series of field trials, Hungarian researchers argued that MON 810 bt. corn introduces many times more Cry toxin proteins into the cultivated area than normal agrotechnology (Székács et al., 2005). Furthermore, they also found that this toxin takes a significantly longer time to decompose in GM crops, in comparison with conventional spraying, due to the toxin being produced by every part of the GM plant that is subsequently ploughed into the soil if it is not harvested. According to the research, they suggested that the persistence might raise questions about possible side effects on soil bacteria, flora or invertebrates. Another study found that the Cry1Ab toxin produced by MON 810 corn caused significant differences in the Mycorrhiza density found in soil (Bakonyi et al., 2006). In a recent study, Lang et al. (2007) found that MON 810 corn has significant impact on biodiversity based on correlations between observed butterfly populations and GM pollen density in living areas. They also estimate that resistance will develop in the target insects within 10 generations.

7.5 Research Objectives

In order to analyze the social discourse of GMOs, we conducted a media study, which allowed us to identify the most influential opinion groups being cited in newspaper articles and news sites. We assumed that different types of mass media play an important role in opinion forming, and at least some of the communications build upon existing and socially robust concerns and prejudices, while others follow the PR-driven line of multinational biotech firms (for different reasons).

In the case of media analysis, the general aim of our research was to produce a quantitative discourse analysis of reports and discussion of GM issues in Hungary, based on Hungarian printed and electronic press in the period June 2000 to November 2008, and with this to contribute to an understanding of how genetic modification plant biotechnology is communicated to the public, as well as uncovering arguments and counter-arguments appearing in the printed and online press.

To understand coverage of genetic modification it is essential to highlight some characteristic features of the Hungarian media. The most important of these are as follows:

(1) In socialist times, the centrally directed press system had been dominated by highly political and propagandist-type newspapers, journals and electronic media. These sources of information were centralized, rigidly following the ideology and policy-line of the state-party (Gulyas 1998).
(2) Since the collapse of the communist system, the number of printed and electronic sources of information has increased drastically. At the same time, there has been a restructuring in media consumption by the population. E.g. The circulation of national dailies declined from 1.69 million in 1988 to 210 thousand in 2008.
(3) There has been a rapid emergence of newly established tabloids, which bear similarities with their Western counterparts, however the social position and quality of tabloid papers are more mixed in terms of social class and income than in Western Europe. People with above average incomes and in higher social positions provide a considerable share of tabloid readership (Gulyas 1998).
(4) The Hungarians are avid television watchers: 98% of households have a television, according to polls 85% of the population watches television on a regular basis. The two major channels are the TV2, whose majority owner is Scandinavian Broadcasting System (SBS) and RTL Klub; owned by a consortium of international investors. The Hungarian Television (MTV) is a public service channel with a marginal audience, has experienced bankruptcy and political scandals. The radio scene is similar to that of television. Older people are the major listeners of public service radio (Lovveless 2008).
(5) The importance of the Internet as a source of information is rapidly increasing, especially in the case of younger and more highly educated people (Dányi and Galácz 2005).
(6) As a consequence of these factors, there is increasing competition between different papers. The most important means of competition is the "tabloidization"

> of sources, focussing on domestic politics, human-interest stories, gossip, and celebrity news. This type of strategy change is typical of the most important Hungarian daily, Népszabadság. This paper was the central official journal of the Hungarian Socialist Workers' Party until 1989 and is often considered to be highly supportive of the Hungarian Socialist Party and the liberal intelligentsia. Foreign media interests own around 74.3% of its shares, especially the Swiss publishing house, Ringier. On the other hand, some papers try to follow a ideologically-driven strategy. The most typical is the Magyar Nemzet (Hungarian Nation), owned by a consortium of Hungarian investors. The paper defines itself as a conservative newspaper. Common themes addressed in the pages of Magyar Nemzet include: demographic and moral decline of the nation; preservation of the historic and traditional culture of Hungary; opposition to cultural and economic globalisation; condemnation of US and Israeli military activities; and support of rights of developing countries to self-determination (Gálik and James 1999).

Foucault's theory understands discourse in terms of a network. Foucault believed that various discourses had the effect of being "monuments... in their own right" (Foucault 1972: 39). A discourse is portrayed as a way of seeing the world composed of values and statements. To Foucault the aim of discourse analysis is to uncover the principles by which statements are dispersed in a discursive field. The definition of a statement is understood in relation to the other statements in the discourse. We can define a statement only by identifying "how it is isolated in the general dispersion of statements" (Foucault 1972: 54).

The tool used to map different statements is called conceptual network mapping (Vedres 2007). A conceptual network is a node graph indicating which statement is connected to others. We have collected all articles from the major Hungarian media outlets that mentioned "genetic modification" for the period in which the research took place. In the first round we separated the discourse centred on genetic modification of plants from other technologies/biotechnologies/modifications (e.g. animal cloning). In the second round we selected articles that contained some valuation and opinion for or against the genetic modification of plants. In the third round we coded the statements. Statements were considered the essential part of individually different utterances. In the first phase we used a relatively high number of categories (e.g. there were separate categories for the statements "genetic modification increases the shelf-life" and "genetic modification facilitates long-range transportation"), then we recoded these statements to form large, robust categories (e.g. collapsing the previously mentioned statements: genetic modification offers logistical advances). The sources of information are summarised in Table 7.3.

In the next phase of research we selected, from a corpus of communications, those sources which could be identified by some institution or organization (leaving out the readers letters, internet-forum remarks, journalists' statements), based only on the last 2 years. Using media analysis, we were able to define the most important opinion groups by their arguments and main messages, in terms of their views on biotechnology, in order to understand their position in the debate (Table 7.4).

Table 7.3 The structure of sources of information for discussion analysis

Source of information	Short characteristics	Number of articles in corpus	Pro-GMO	Anti-GMO
Népszabadság	Left-liberal newspaper	51	43	8
Magyar Nemzet	Conservative newspaper	32	12	20
Élet és Irodalom	Liberal weekly newspaper	12	12	0
Magyar Hírlap	Conservative newspaper	24	7	17
Index.hu	One of the leading online-only Hungarian language news Portals	9	0	9
Origo.hu	One of the leading online-only Hungarian language news Portals	7	0	7
Other	Different printed press archives	12	7	5
Magyar Tudomány	Central journal of Hungarian Academy of Sciences	8	4	4
Forumkereso.hu	Collection of online-discussion fori	134	23	111
Total		289	108	181

We decided to describe each group on the basis of their relationships to the following key questions:

1. *How do you see the implementation of the GMO regulation?*
2. *Is there any present need for GM crops and food in Hungary?*
3. *How do you judge sustainability of the coexistence of traditional and GM species.*
4. *For which social and economic actors could GM technology offer benefits?*
5. *How would the permission of GM crop production affect the international competitiveness of Hungarian seed production?*
6. *What – if any – essential differences are there between traditional plant breeding and GM technology?*
7. *Do you anticipate that any ecological risks will arise from GM crop production?*
8. *Are consumers guaranteed freedom of choice? (In other words: Do you believe that labels tell the truth? How effective does product traceability work in this field? Are national food safety authorities strong enough to regulate it?)*
9. *Are there any risks that arise from organic farming?*
10. *Do you believe that GM technology could ameliorate famine (now or in the future)?*
11. *What is your opinion of "GM-free regions"?*
12. *Do you have any concerns about possible side effects (of GM) on human health?*

Table 7.4 Opinion groups (identified on similar communication panel usage) and their absolute and relative share in media coverage weight

Opinion group Biotech industry	Absolute share	Relative weight %
Monsanto Trading Ltd	11	12.94
	–	**12.94**
Institutes of authority		
Ministry of Agriculture and Rural Development	10	11.76
Ministry for Environment and Water	6	7.06
Hungarian Office for Food Safety	6	7.06
National Institution for Food Safety and Nutritional Sciences	1	1.18
	–	**27.06**
Environmental organisations	9	**10.59**
Greenpeace Hungary	9	10.59
Ökotárs Foundation – environmental group	3	3.53
Élőlánc (environmental political formation)	3	3.53
International environmental organisations	5	5.88
	–	**23.53**
Academic institutions I.		
Institution for Plant Protection of the Hungarian Academy of Sciences	4	4.71
Association of Hungarian Plant Breeders	1	1.18
	–	**5.88**
Academic institutions II.		
Biological Research Centre of the Hungarian Academy of Sciences	3	3.53
Agricultural Research Institute of the Hungarian Academy of Sciences	2	2.35
Agricultural Biotechnology Centre	2	2.35
	–	**8.24**
Agribusiness sector		
Organic farmers and their associations	2	2.35
Association of Hungarian Farmer Groups and Organizations	1	1.18
Hungarian Association of Crop Processors, Feed Manufacturers and Traders	1	1.18
Association and Chamber of Seed Producers	1	1.18
	–	**5.88**
Consumer representatives and consumer studies		
Authority for Consumer Protection	1	1.18
Hungarian Association for Consumer Protection (NGO)	1	1.18
	–	**2.35**
The seven groups cover 85.88% of all media communications		

In the final phase of the survey we contacted at least one (but preferably more) of the dominant actors in each group and – through interviews – asked them to comment on, complete or correct the characterization of their own groups.

We found that different opinion groups work within the argumentation frameworks of the same organizations. The most notable exception to this was the case of academic institutions (listed in Table 7.4 as Academic institutions I and Academic institutions II) where their arguments show a clear divergence.

The group of agricultural producers has formed a relatively homogenous group in their opinions of GM. The goals of agricultural producers were simple: to grow marketable products. The social and environmental aspects of this question were only marginal problems for them. From the point of view of agricultural producers there were no separable opinion-differences. Their basic goal was producing goods they could sell.

7.6 Results of Media Analysis

As a general tendency, it can be stated that the GM discussion received relatively little coverage in the Hungarian media. In numerous cases, some charismatic figures played a key role in determining the focal points of the debate. It can be seen from Table 7.3 that there are characteristic differences between the different groups on the GM debate.

Our interpretation of the news mirrors the general line of the papers. The overwhelming majority of articles published in Népszabadság supported genetic modification. When news could be interpreted as anti-GM, they tried to decrease its importance. E.g. When the Népszabadság reported the results of research at Vienna University, which found that third or fourth generation mice fed on a diet of genetically modified corn had significantly lower weights than the preceding generations and a decrease in fertility, the newspaper editors framed the story with the title: "Eat GM corn – you won't need a condom!" (Népszabadság, 13.11.2008). A day later, the paper's scientific editor went on to write: (1) that the citation had been taken from Greenpeace, and so it was not a proper scientific study; (2) the (potentially) negative effect in corn does not apply to other foods (e.g. soy); (3) some GM crops (e.g. golden rice) are able to produce vitamin A-rich nutrition; and (4) in the end, decreasing fertility would be a much more humane way of regulating population growth, than suffering high levels of infant mortality (Népszabadság, 13.11.2008). At first sight this seems to be a rather confusing argument, but – in our experience – it is indicative of the level of scientific communication in the most important Hungarian newspaper.

Another typical example is when the Népszabadság published a short article on the rapid spread of GM crop cultivation worldwide. The style of the article was neutral, but it was followed by a quotation from the Minister of Agriculture and Rural Development on the need for an increase in agricultural production in Hungary. The article ended with the suggestion that the Minister supports GM technology as a means of increasing production (Népszabadság, 23. 01. 2007).

This is in line with the observation made by Cook et al. (2005: 7), when they analyzed the practices of the British press in GM related news: "Some newspaper articles state facts, typically in final position, without making the connection to preceding text explicit, thus leaving relevance to be inferred."

The conservative newspaper Magyar Nemzet, which supports the political opposition, has published articles and letters criticizing genetic modification, highlighting the adverse effect of this technology on small-scale farms. At the same time, pro-GM articles have been published. One of these, written by a member of the Hungarian Academy of Sciences, highlighted the importance of genetic modification at a time when "hunger riots" could take place in Hungary as a result of the "anti-human" politics of the government (Magyar Nemzet 22.05.2008). This is a striking example of how pro-GM arguments and political propaganda are joined together.

Interestingly, websites of civil organizations concerned with GMOs are poor; e.g. the website for the Hungarian branch of Greenpeace (greenpeace.hu) offers only some general and simplistic counter-arguments about GM. At the same time, it claims that the "Hungarian government has abandoned its decision to allow the growing of GM-crops, at Greenpeace's request". However, as we have seen this is a much more complex problem. The most important arguments, along with their frequency, are summarized in Table 7.5.

The set of counter-arguments against GM crops is heterogeneous. The most important of these are summarized in Table 7.6.

The networks of arguments for application of genetic modification are depicted in Fig. 7.1.

The size of the nodes (circles) is approximately proportional to their frequency. The lines between the nodes show the co-occurrence of different arguments in one communication; their thickness is proportional with the frequency of joint use of different arguments (Borgatti et al., 2002). It is obvious that the most important

Table 7.5 Arguments in the GM debate

Arguments and their acronyms		Relative frequency of occurrence
1.	Reduction of chemical use (NOCHEM)	15
2.	Increasing of production (PROD)	11
3.	Solution of food supply for Third World countries (FOODSEC)	17
4.	Improvement of nutritional value functional properties (e.g. increased vitamin content) (FUNCTIONAL)	6
5.	Increased stability of production (STAB)	12
6.	Higher incomes for farmers (INOME)	8
7.	Logistical advances (LOGIST)	5
8.	No evidence for harmful effects (NOEVID)	5
9.	Innovative, new technology, the legal regulation of which goes against the principle of economic freedom (contra-Luddite approach, historical parallels) (INNOV)	20
10.	Material for bio-fuel (BIOFUEL)	1

Table 7.6 Counter-arguments in GM debate

Arguments and their acronyms	Relative frequency of occurrence
Need for more information (INFO)	17
Allergenic problems (ALLERG)	11
GM-resistant weeds and pests (SUPERWEED)	4
Uncontrolled cross-pollination (CROSSPOLL)	6
Low profitability (LOWPROF)	5
Threat for bio-production (BIOPROD)	8
Threat for small-scale farms (SMALLFARM)	7
Possibility of biotech companies abusing economic superiority in third world (ANTIMONOP)	19
Marketing problems of GM products (MARKETPROB)	6
No need for more production (NOTNEC)	8
Higher food prices (HIGHPRICE)	6
Change of Nature (NATURE)	3

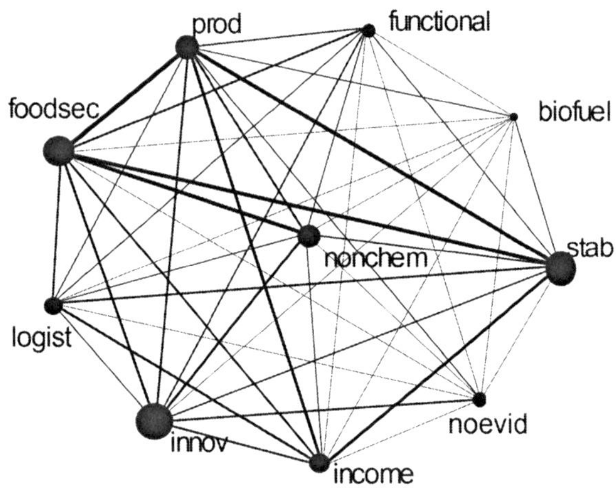

Fig. 7.1 Network of pro-GM arguments

arguments in pro-GM communications were the high innovation content of the technology, food security and the increase in production stability. The latter is especially important in Hungary, where drought is a key problem in agricultural production. The emphasis on the food security argument is less striking, as the attention of Hungarian citizens is focussed mainly on domestic problems (e.g., in 2007, 81% of front page headlines in Népszabadság concerned domestic issues).

The arguments against genetic modification (Fig. 7.2) show a more dispersed picture. In this set of arguments the main emphasis is on anti-monopolistic attitudes and the need for more information on the effects of genetic modifications. Cook et al. (2005) similarly found that the anti-GM press tends to site agricultural biotechnology in a more global frame.

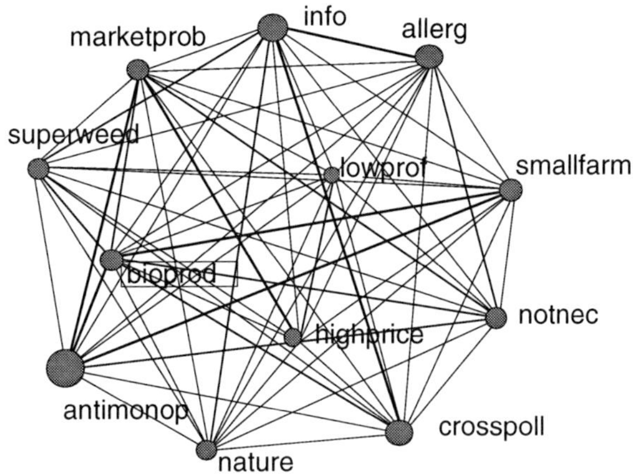

Fig. 7.2 Network of anti-GM arguments

7.7 Accounts of the Opinion Groups

7.7.1 Biotech Industry

Industry representatives argue that there is a need for GM crops in Hungary, which could significantly improve farmer livelihoods. Direct and indirect advantages are both tangible and intangible. The direct advantage is that farmers could achieve higher yields and cost reduction. The indirect advantage is that society in general could benefit from environmentally friendly production technologies and reduced chemical residues in food products. The follow-up interviews aimed at company leaders strengthened this finding.

7.7.2 Environmental Organizations

Our interviewees working in environmental organizations strongly opposed the introduction of the first generation of GM crops, believing that biotech companies are the exclusive beneficiaries. In the follow-up interviews they seemed to be more concerned about the possible economic consequences rather than the ecological side effects (although many of their written communications, e.g. their websites, focused on ecological problems). They emphasize the presumptive disadvantage for Hungarian seed and crop production, according to output markets, which now prefer products of Hungarian origin because of their GM-free attributes. Regarding consumers, they argue that no benefit can be delivered to them via the present GM technology.

As we have shown earlier in this chapter, socio-economic development in Hungary has not promoted the emergence and increasing public acceptance of organic (bottom-up) development of these organizations. That is why we have to take into consideration the possibility that the opinions expressed by environmental organizations are shared only by a rather limited number of activists, or are simply reproductions of the views of their parent organizations.

7.7.3 Agribusiness Sector

Representatives of agribusiness argue that GM crops are practically useless under present Hungarian conditions, as there is no added value beyond normal hybrids. According to them, the modified features of GM applications provide protection against insects that pose insignificant threats to Hungarian crops. They would accept more readily a type of biotech crop that would resist more malignant pests. There is a segment that would adopt GM if it were legal and would offer economic advantages. Hungarian agricultural producers' approach has numerous parallel features with that of other states. Yamaguchi and Harris (2004), in analyzing the Bt cotton discourse in India, found that the dominant frame has shifted over time from governmental process to economic impact. The real economic analysis of the impact of GM technologies is hindered by the fact that in Hungary there is no reliable economic information system on cost-benefit analysis of different agricultural products, thus, there is the possibility that economic data and calculations could be manipulated.

Some others (especially organic farmers) worry about coexistence in the Hungarian context, which they believe is not sufficient (in spite of it being very strict compared to other EU countries), and as such, may cause serious damage to export opportunities.

7.7.4 Academic Institutions I (Anti-Biotech Academic Sphere)

Social aspects of biotechnology lie outside the main research agenda of Hungarian social scientists. This can be explained by three factors: (1) the social situation (e.g. rapid social changes during transition, formation of new elites, increasing poverty, social prejudices and integration of minorities) are much more acute problems than biotechnology; (2) the social aspects of technology are considered technical rather than social problems, (3) in general, there is a wide gap between "social scientists" and "natural scientists". Papp (2002) analysed the content of 100 Hungarian sociology articles in the 1990s and found only one on social aspects of science and technology.

The first group of academic institutions believe that the introduction of GM crops would mostly benefit biotech companies. Their view is that the separation of GM and non-GM would be impossible to implement, and therefore coexistence in the Hungarian context would be uncontrollable in the way that it has been proposed. They believe that exports of Hungarian crops would suffer after the introduction of GM technology. They maintain that guaranteeing GMO traceability is already impossible. For them, field trials play an important role and they categorically oppose research results being influenced by multinational companies in any way.

7.7.5 Academic Institutions II. (Pro-Biotech Academic Sphere)

These institutions present a contrasting stance to Academic institutions I. They emphasize the importance of Hungary joining GM supporter countries as soon as possible in order to gain competitive advantage over other European countries. They argue that behaving in this way would enable the country to be among the greatest seed exporters once again.

They call GM technology the cutting edge in R&D, and consider resistance to it as "economic suicide", because it will ultimately spread through Europe as it did many other places throughout the world. They believe that advantages, such as better quality or higher nutritional value of the second and third generations of GM plants, will convince those who still resist. A twist in their strategy is to focus on non-food GM crops, anticipating less reluctance by consumers. They also believe that people will recognize GM energy crops as being a genuinely green (environmentally friendly) technology. Most of these institutions have already signed research contracts with biotech companies. They regard these partnerships as the only chance Hungarian agricultural R&D has to survive, using the century-long traditions of plant selection based on a highly qualified labour force and production culture. On the possibility that GM presents marketing disadvantages, these academic institutions have surmised that it only accounts for the short term. Once Europe's market becomes liberalized people will learn that GM technology offers no more risk than conventional agriculture.

In our estimation, the majority of Hungarian scientists concerned (approximately three-quarters of them) adopt – at least officially – a pro-GM attitude. Behind this, a social-psychological aspect may be presumed: they do not want to show themselves as "conservative" or "isolationist" even if they cannot prove scientifically that GM is harmless.

7.7.6 Governmental Institutions

Officers of governmental institutions were not forthcoming with their views. We could find definite opinions only very rarely, if at all. Nonetheless, comments revealed certain concerns from the perspective of the authorities:

- Economic (coexistence, possible reduction of export potential)
- Ecological (gene leaking, biodiversity)
- Controllability (keeping limit values, detectability, consumer choices)

7.7.7 Consumer Representatives and Consumer Studies

In Hungary, consumer groups are less prominent than environmental groups. Consumers are therefore often represented by environmental NGOs or the state itself by its official consumer protection authority. Another way to get consumer accounts is to conduct representative opinion polls and other surveys. These primary research methods (Lakner et al., 2006) help us to understand consumer behaviour, and include mappings of food risk perception between 2005 and 2007 in cases like BSE, avian influenza, GM food, food additives and food counterfeits. In our questionnaire-based survey for this study, we aimed for representativeness in age and gender (and if possible, size of settlement and geographical dispersion, as well) and a sample size of approximately 1000 respondents.

Our actual survey in 2007 shows that attitudes to GM food have been polarized compared to data from 2001 (Figs. 7.3 and 7.4). The segments representing people with a positive approach seem to have decreased during this 6-year period.

We asked participants to compare different (and very often contested) technologies and procedures in order to ascertain comparable acceptability. As Fig. 7.5 shows, GM food is less accepted than nuclear energy and genetic testing carried out by insurance companies to determine coverage. However, genetically modified organisms have been extensively used in food processing for the last 2 decades and yet, they still seem not to be accepted by respondents, suggesting that people may not know they are eating foods with GM ingredients.

Figure 7.6 presents the order of the different technologies aligned by the relative standard deviation [(standard deviation of X)/(average of X)] of the consumer judgments. While all types of cloning and genetically modified food also finished at the top of the list (disclosing no consensus on behalf of the respondents), evaluation of

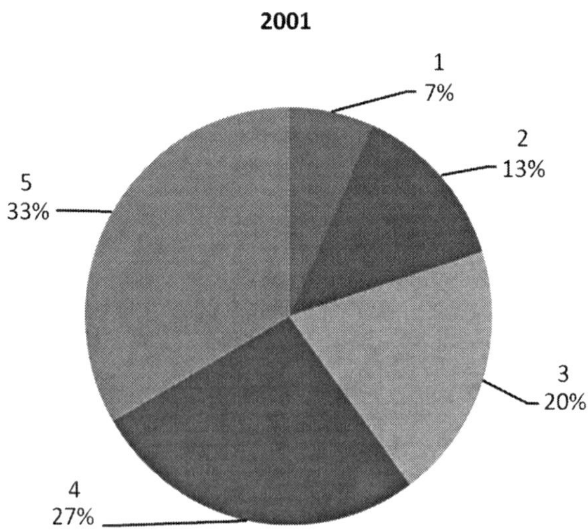

Fig. 7.3 Attitudes toward genetically modified food (2001). *1*: Strongly positive, *2*: Mostly positive, *3*: Neutral or mixed, *4*: Mostly negative, *5*: Strongly negative

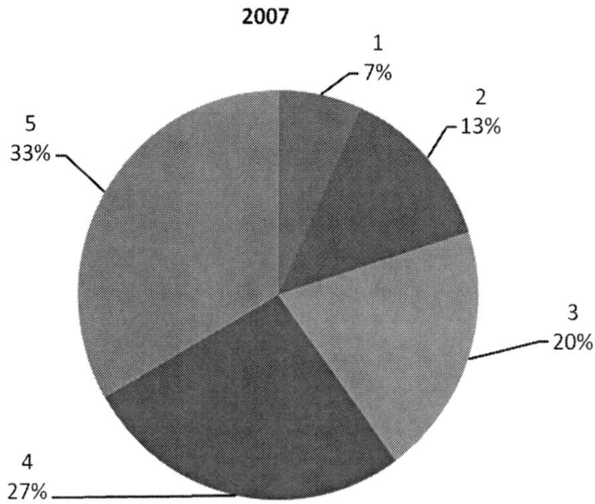

Fig. 7.4 Attitudes toward genetically modified food (2007). *1*: Strongly positive, *2*: Mostly positive, *3*: Neutral or mixed, *4*: Mostly negative, *5*: Strongly negative

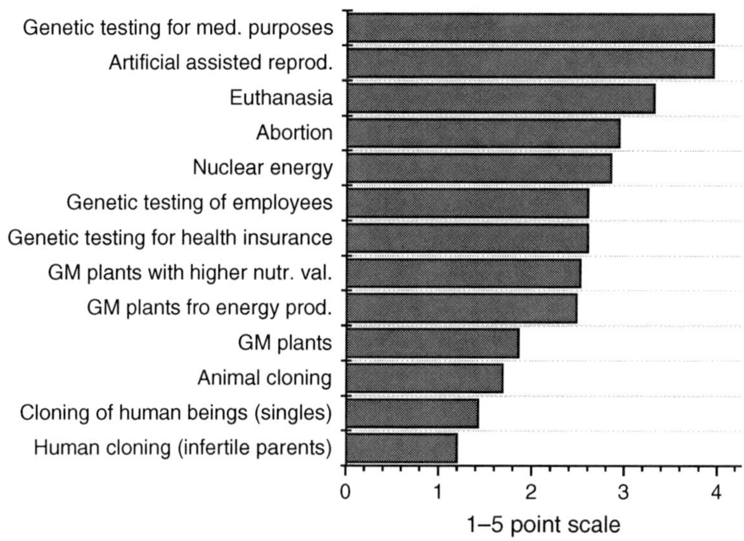

Fig. 7.5 Acceptance of different technologies and procedures (means)

"older" technologies and procedures, like assisted fertilization, abortion, euthanasia and nuclear energy seem to enjoy a more established common approach.

7.8 A Contradictory System

Many different approaches can be observed with regard to social resonances of food oriented GM technology. The majority of opinions expressed in the interviews are driven by various concerns leading to robust social debate. There are a minority of

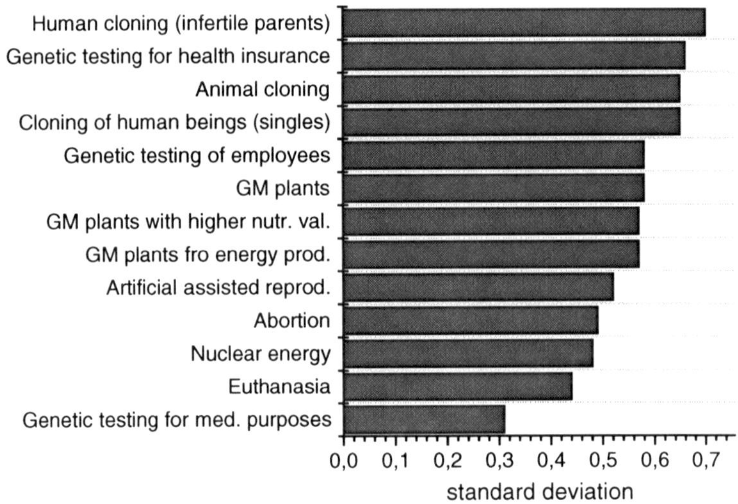

Fig. 7.6 Relative acceptance of different technologies and procedures

well-articulated voices, at the same time. Biotech companies (most often Monsanto) and academic biotech research institutions continue an agile and sophisticated communication strategy. They use print, broadcast and online media confidently, and they seek opportunities for novel forms of science public relations and public education (for instance, they actively participated in the successful popular science series called ENCOMPASS or University of All Knowledge (Rédey 2006), they organize events promoting their achievements and have a positive attitude to giving media or social science interviews). All these efforts increasingly lead them to define the frames of public discourse on genetic engineering. Arguments of biotech firms and scientists in Hungary focus on the potential of GM products as the solution to malnutrition, and the importance of GM crops in an era of climate change. This is an important problem in Hungary, due to the increasing incidence of drought in the country. The environmental risks of GM crops appear relatively less important in communications, possibly because these aspects seemed not to have a direct effect on the different stakeholders.

This is a natural phenomenon, because these questions are less understandable for the wider audience. In communications concerning the possible human health-related effects, the main emphasis is on freedom of choice for consumers. This argument is in line with the spirit of the post-transition age: after years of paternalistic socialism (Swain 1992) when the citizens' decisions were considerably bound by the state, the market economy offered *choice*.

The concerns of anti-biotech commentators that Eastern countries would become a "Trojan horse" allowing GM technology into Europe through the back door proved to be mostly false. However, there are still some pitfalls on the battlefield. A number of institutions belonging to the Hungarian Academy of Sciences broadly support GM technology. The Academy tries to be neutral (at least formally), but some of its scientific institutions are in close co-operation with multinational biotech firms, doing contracted research. The amount of income from companies for this activity is hard to quantify, as it is confidential data. In some specialists' opinion, there is considerable discrepancy between the "official" and the "real" research being carried out. For example, the head of a scientific laboratory has stated his view on the official homepage of the National Institute of Research and Technology: "I can cite a lot of examples where a research institute is officially carrying out 'biotech research', but in practical terms they are just doing simple and routine measurements for international biotech firms" (Szabó 2005). According to estimates by different experts (asking to remain anonymous), approximately 40–60% of the budgets of some academic biotech institutes are covered by contracted research activities to firms. In this way, such firms could theoretically influence these institutions and, therefore there is a potential conflict of interest. Through the active influence of the scientific community and policy makers, GM could be introduced into the EU through the backdoor of Hungary. The reasons for this include strong existential pressure, aggravated by decreasing state subsidies for science and technology research (which now amounts to less than 1% of GDP). Ironically, such institutions were primarily organized to use governmental funds to ensure the *independence of science*. R&D contracts with biotech companies are undoubtedly lifesaving in the short-term, and may become a strategic asset in the long run, because Hungarian scientific institutions and university departments obtain only half of their budget from the state, the other half is supposed to come from business or other sources (e.g. National or EU research funds). It seems, though, that investment in science comes with certain conditions (at least when it comes to technology promotion).

There are numerous contradictions in the declarations of the Hungarian political elite on the subject of genetic modification. All Hungarian parties supported the national GM moratorium in January of 2006. At the same time, the Hungarian MEP Béla Glattfelder (member of the EP Committee on Agriculture and Rural Development), made a definite pro-GM statement at a conference that was financed by biotech companies and took place in the Hungarian Academy of Sciences (Glattfelder 2006). He suggested 14 amendments to the Committee on International Trade for the European Parliament's report (2006) that were in favour of dissemination of GM crops. These include statements like: "... a total ban on GMO crops is not a viable option"; "regrets that as a consequence of the restrictive policy and regulatory approach applied by the Commission and some Member States, the European biotechnology sector is at risk of lagging behind" and "calls on the Commission and the Member States to keep Community and national legislation in line with WTO rules and obligations" (Opinion of the Committee on International Trade 2006). Comparing these two declarations, the contradiction is obvious. In the first case ("for the domestic audience") the opinion of the party politician is closer to an anti-GM approach in the second case (mainly for the business sphere and international audience) the opinion supports pro-GM interest groups.

There are some other aspects to the food production and trade question in relation to GMOs and Hungary. There are rumours that some farmers have used GM seeds in commercial agricultural production. In spite of the effort we made to identify such fields, we have not yet found sound evidence that this takes place. Our enquiry (Sebestyén 2008) has revealed that while there has been regular monitoring of GMOs in food production, GMO content in unlabelled foods regularly exceeds the legal 0.9% limit by 2–7%, while another 35–60% of the samples analyzed also contain GM ingredients, although below legal limits. This basically means that GMOs have become a food that Hungarians consume on a daily basis, including those who wish to avoid eating GM food. Hungarian society seems to have no choice other than to accept that it is technically challenging to produce food with 0% GMO content when imported soy from overseas is used. Two independent sources doubted that GM-free soy could be guaranteed at all, because of improper transportation methods. This may mean that consumers will need to get used to it step by step over time.

7.9 Conclusions

Einsiedel et al. (2001) report on the success of consensus conferences in Denmark, Canada and Australia for decision making on the GMO issue. Results of our research suggest that there may only be limited possibilities for consensus building through social debate on the GM issue in Hungary, because there is a lack of transparency of interests and roles.

Furthermore, the academic sphere is divided: there are pro- and anti-biotech groups, which confuses and makes it difficult to achieve an "independent scientific opinion". At present, there is an inherent contradiction in the positions of academic organizations in Hungary: on the one hand, society in general expects them to fulfil the role of "scientific watchdog", but on the other, the state of affairs means that scientific debate is contradictory, and arguments are mixed with emotion. Political parties tend to be opportunistic, trying to maximize support in different political environments.

Hungarian agriculture has not been able to develop a coherent, widely accepted agricultural strategy. GM crops are useful mainly for mass production, large-scale farms (Gray 2004), but in the case of Hungary it is an open ended question; which should be preferred by agricultural policy: the small- to middle-scale, family-based farms (similar to most Western European states), or the large-scale commercial farms owned by economic enterprises, such as in the eastern part of Germany? Under these conditions the state organs cannot follow a straightforward regulatory path.

Being GM-free for as long as possible or being a GM producer country both offers tangible advantages and poses threats. Without taking a stand on either side, it is apparent that there is widespread public rejection of GM food and first generation GM crops, which emerges from scientifically established arguments, moral reasons and media driven perceptions.

There are well-defined opinions on genetic modification, but in some cases it is difficult to determine the real position of some stakeholders (e.g. political parties) in the debate. It is likely that local processes and global trends (such as pressures exerted by the WTO) will bring about GMO market liberalization; further research is necessary to investigate how this might affect public attitudes.

References

Ágh, A. 2001. Early democratic consolidation in Hungary and the Europeanisation of the Hungarian polity. In *Prospects for democratic consolidation in East Central Europe*, eds. G. Pridham and A. Ágh, 270. Manchester, NY: Manchester University Press.

Angelusz, R., and R. Tardos. 2001. Change and stability in social network resources: The case of Hungary under transformation. HTML. In *Social capital: Theory and research*, eds. N. Lin, R. Burt, and K. Cook, 297–323. New York, NY: Aldine de Gruyter.

Bakonyi, G., F. Szira, I. Kiss, I. Villányi, A. Seres, and A. Székács. 2006. Preference testswith collembolas on isogenic and Bt-maize. *European Journal of Soil Biology* 42:132–135.

Bánáti, D., and G. Kasza. 2003. In *Biotechnológia: A magyar fogyasztók és szakemberek élelmiszer-biztonsági kockázatészlelése,* ed. T. Júlia, 103–108. Budapest: Business Class Kiadó.

Bánáti, D., J. Popp, and N. Potori. 2007. A GM növények egyes szabályozási és közgazdasági kérdései. Agrárgazdasági tanulmányok, 200703, 138.

Beck, U. 1992. *Risk society: Towards a new modernity*, 436. New Delhi: Sage.

Belt, H., and J. Keulartz. 2007. Worldwide cultural diffeneces in socio-etical views in relation to biotechnology. A report commissioned by the Netherlands Commission of Genetic Modification, 210.

Birner, R., and G. Alcaraz. 2004. Policy dialogues on genetically modifies crops in Europe: Insights for African policy dialogues on biotechnology, A background paper prepared for the second session of the African policy dialogues on biotechnology-Southern Africa www.ifpri.orgafricadialoguepdfpolicydialoguspaper.pdf. Accessed 08 Apr 2008.

Borgatti, P., M. Everett, and L. Freeman. 2002. UCINet 6 Network Analysis Software. AnalyticTechnologies.

Bryner, G. 2001. Cooperative instruments and policy making: Participation in US environmental regulation. *European Enviornment* 11(1):49–60.

Committee on International Trade. 2006/2059. For the Committee on Agriculture and Rural Development on Biotechnology: Prospects and challenges for agriculture in Europe, 164.

Cook, G., P.T. Robbins, and E. Pieri. 2005. "Words of mass destruction": British newspaper coverage of the genetically modified food debate, expert and non-expert reactions. *Public Understanding of Science* 14(1):1–25.

Curtis, K.R., J.J. McCluskey, and T.I, Wahl. 2004. Consumer acceptance of genetically modified food products in the developing world. *AgBioForum* 7:1–2.

Dányi, E., and A. Galácz. 2005. Internet and elections: Changing political strategies and citizen tactics in Hungary. *Information Policy* 10(3–4):215–227.

de Marchi, B., and J. Ravetz. 2001. Participatory approaches to environmental policy. Environmental Valuation in European Policy research brief, Cambridge Research of the Environment, No. 10, 146.

Einsiedel, E.F., E., Jelsoe, and T., Breck. 2001. Publics at the technology table: The consensus conference in Denmark, Canada, and Australia. *Public understanding of science* 10(1): 83–98.

Eurobarometer. 2006. Europeans and Biotechnology in 2005: Patterns and Trends; Full Report, 23.

European Parliament, Committee on International Trade. 2006. Biotechnology: Prospects and challenges for agriculture in Europe (20062059(INI)) Amendments 1-50, Draft opinion – Jonas Sjöstedt (PE 374.050v01–00).

Foucault, M. 1972. *The archeology of knowledge*, 239. London: Tavistock.

Frewer, L., J. Lassen, B. Kettlitz, J. Scholderer, V. Beekman, and K.G. Berdaf. 2004. Societal aspects of genetically modified foods. *Foods and Chemical Toxicology* 42(8): 1181–1193.

Gálik, M., and B. James. 1999. Ownership and control of the Hungarian press. *Javnost-The Public* 6(2):75–92.

Gaskell, G., N. Allum, M. Bauer, J. Durant, A., Allansdottir, H., Bonfadelli, D., Boy, S., de Cheveigné, J.B., Fjaestad Gutteling, J., Hampel, E., Jelsøe, J.J., Correia, M., Kohring, N., Kronberger, C., Midden, T.H., Nielsen, A., Przestalski, T., Rusanen, G., Sakellaris, H., Torgersen, T., Twardowski, and W., Wagner. 2000. Biotechnology and the European Public. *Nature Biotechnology* 18(6):935–938.

Gaskell, G., N.C., Allum, and S.R., Stares. 2003. Europeans and biotechnology in 2002: Eurobarometer 58.0. European Commission, 180.

Gaskell, G., M., Bauer, and J., Durant. 1998. Public perceptions on biotechnology in 1996: Eurobarometer, 46.1. In *Biotechnology in the public sphere: A European sourcebook*, eds. J. Durant, M. Bauer, and G. Gaskell , 189–214. London: Science Museum.

Glattfelder, B. 2006. Genetikailag módosított termékek nemzetközi szabályozása (oral presentation). Hungarian Academy of Sciences, Department of Agricultural Sciences – Pannon Associaton for Plant Biotechnology. 27. 04. 2006, Hungarian Academy of Sciences, 120.

Gray, A. 2004. Genetically modified crops: Broader environmental issues. *Journal of Commercial Biotechnology* 10(3):234–240.

Gross, A. 1994. The roles rhetoric int he public understanding of scinece. *Public understandin of science* 3(1):3–23.

Gulyas, A. 1998. Tabloid newspapers in Post-Communist Hungary. *Javnost-The Public* 5(1): 65–77. http://www.ejc.net/media_landscape/article/hungary/

International Herald Tribune. 2008. http://www.iht.comarticles20080326businessorganic.php Romania reconsiders its welcome of biotech corn by James Kanter. Published: 26.03.2008.

Irvin, R. A., and J., Stansbury. 2004. Citizen participation in decision making: Is it worth the effort? *Public Administration Review* 64(1):55–65.

Izsáki, Z. 2004. *Szántóföldi növények vetőmagtermesztése és kereskedelme*, 27–32. Budapest: Mezőgazdasági Kiadó, http://www.tankonyvtar.hu/main.php&objectID=5335129. Accessed 08 Dec 2008.

Janasoff, S. 2005. *Designs on nature: Science and democracy in Europe and the United States*, 510. Princeton, NJ: Princeton University Press.

Kornai, J. 1992. *The socialist system: The political economy of communism*, 736. Princeton, NJ: Princeton University Press.

Lakner, Z., and I., Hajdu (2004): Possibilities and problems of innovation in the Hungarian food industry. Studies in Agricultural Economics No.100, 132.

Lakner, Z. I., Hajdu, D., Bánáti, E., Szabó, and G., Kasza. 2006. The application of multivariate statistical methods for understanding food consumer behaviour. Studies in Agricultural Economics 2006 No. 105, 645–672.

Lakner, Z., and G., Kasza. 2005. Hungarian consumers and genetic engineering: what's behind? *Technology and Society* 15(2):6–8.

Lang, A., E., Lauber, and B., Darvas. 2007. Early-tier tests insufficient for GMO risk assessment. *Nature Biotechnology* 25(1):35–36.

Lemaux, P.G. 2008. Genetically engineered plants and foods: A scientist's Analysis of the issues. *Annual Review of Plant Biology* 59: 771–812.

Lovveless, M. 2008. Media dependency: Mass media as sources of information in the democratizing countries of Central and Eastern Europe. *Democratization* 15(1):162–183.

Macnaghten, P., M.B., Kearnes, and B., Wynne. 2005. Nanotechnology, governance, and public deliberation: What role for the social sciences? *Science Communication* 27(2):268–291.

Malthus, T.R. 1820. *Principles of political economy considered with a view to their practical applications*, 386. London: Longman, Hurst, Rees, Orme and Brown.

Marris, C. 2001. Purblic views on GMOs: Deconstructing the myths. *European Molecular Biology Organisation reports* 2(8):542–548.
Mccombs, D.L., R.T., Shaw, and H.L., Weaver. 1997. *Communication and democracy: Exploring the intellectual frontiers in agenda-setting theory*, 358. Mahwah, NJ: Lawrence Erlbaum.
Meadows, D.H., D.L., Meadows, J., Randers, and W.W., Behrens. 1972. *The limits to growth*, 388. New York, NY: Universe Books.
Nap, J.-P., P., Metz, M., Escaler, and A.J., Conner. 2003. The release of genetically modified crops into the environment Part I: Overview of current status and regulations. *The Plant Journal* 33(1):1–18.
Nisbet, M.C. 2006. Attention cycles and Frames in the Plant Biotechnology Dabate. *The Harvard International Journal of Press Politics* 11(2):3–40.
Nowotny, H., P., Scott, and M., Gibbons. 2001. *Re-thinking science: Knowledge and the public in an age of uncertainty*, 266. London: Polity Press.
OECD-FAO. 2008. The OECD-FAO Agricultural Outlook 2008–2017.
Paarlberg, R. 2003. The global food fight. In *The cultural politics of food and eating*, eds. J.I. Watson and M.I. Caldwell, 276–286. London: Blackwell.
Papp, Z.A. 2002. The down of metaphors? Sociological discourses in nineties in Hungary. In *What's the news in Central –Eastern European sociology?* ed. É. Kovács, 71–105. Budapest: Teleki László Foundation.
Pickvance, C.G. 1999. Democratisation and the decline of social movements: The effects of regime change on collective action in Eastern Europe, Southern Europe and Latin America. *Sociology* 33(2):353–372.
Priest, S.H. and Gillespie, A.W. 2000. Seeds of discontent: Expert opinion, mass media messages, and the public image of agricultural biotechnology. *Science and Engineering Ethics* 6(4): 529–539.
Rédey, S. 2006. Communication technologies of science – a change of science-cognition. 5th Annual IAS-STS Conference: Critical Issues in Science and Technology Studies.
Santaniello, V., R.E., Everson, and D., Zilberman. 2001. *Market development for genetically modified foods*, 388. Wallingford: CABI publishing.
Sebestyén, E. 2008. Genetikailag módosított összetevők jelenléte a magyar élelmiszer- és takarmány előállításban. Thesis, Budapest, Corvinus University of Budapest.
Slovic, P. 1987. Perception of Risk. *Science* 236:280–285.
Slovic, P., B., Fischhoff, and S., Lichenstein. 1985. Characterising perceived risk. In *Perilous progress: Managing the hazards of technology*, ed. R.W. Kates, 266. Boulder, CO: Westview Press.
Spicer, A., G.A., McDermott, and B., Kogut. 2000. Entrepreneurship and privatisation in Central Europe: The tenuous balance between destruction and creation. *Academy of management Review* 25(3):630–649.
Swain, N. 1992. *Hungary: The rise and fall of feasible socialism*, 322. Budapest: Verso.
Szabó, M. 2005. Comment to the dispute on role of the biotechnology in the national economy, http://www.nkth.gov.hu/portalforum/lehet-huzoagazat-080519. Accessed 05 Feb 2009.
Székács, A., J., Juracsek, L.A., Polgár, and B., Darvas. 2005. Levels of expressed Cry1Ab toxin in genetically modified corn DK-440-BTY (YIELDGARD) and stubble. *Federation of Biochemical Societies Journal* 272:s1.
Vedres, B. 2007. Pathways from postsocialism: Ownership sequence and performance of firms in Hungary, 1991–1999. *European Management Review* 4(2):93–105.
Vilella-Villa, M. and J., Costa-Front. 2008. Press media reporting effects on risk perceptions and attitudes towards genetically modified (GM) food. *Journal of Socio-Economics* 37(5): 2095–2106.
Wedel, J.R. 2003. Clans, cliques and captured states: Rethinking "transition" in Central and Eastern Europe and former Soviet Union. *Journal of International Development* 15(4):427–440.

Wynne, B. 2006. Public engagement as a means of restoring public trust in science – Hitting the notes, but missing the music? *Community* 9(3):211–220.

Wynne, B. 2007. Taking European Knowledge Society Seriously – Report of the Expert Group on Science and Governance to the Science, Economy and Society Directorate, Directorate-General for Research, European Commission, 188.

Yamaguchi, T. and C.K., Harris. 2004. The economic hegemonization of Bt cotton discourse in India. *Discourse & Society* 15(4):467–491.

Chapter 8
Coping Strategies and System Adaptation of Agricultural Biotechnology Research in Hungary

Farah Huzair

8.1 Introduction

Initial reforms in Hungary took place in the 1960s with the aim of transition to a market economy. The country then underwent a significant program of more intensive reforms in the early 90s when Soviet rule ended. Austerity measures in line with IMF guidelines were introduced in 1995 and included expenditure cuts, limits in public sector wage increases, a devaluation of the forint against the dollar, and an increase in import duties (Jeffries 2002). Economic history seriously effected development of the agri-biotech innovation system. Amidst the economic crisis, science and development was not at the top of policy makers agendas (Chataway 1999). It has been argued that in an attempt to reverse the extremely interventionalist role of the state during the socialist regime, the draw back of the state from many policy areas resulted in a transitionary period where the state did not intervene enough (Von Tunzelman 2005). In addition to this, the Hungarian biotechnology sector has also to contend with complications in the regulation of GM crops arising from the strongly restrictive national legislation adopted by Hungary under the EU overarching framework. The basis for adoption of this national legislation is highly contested. These economic and political factors together, raise questions about the development of science in the context of an uncertain political climate and contested regulatory framework. This chapter presents original qualitative ethnographic data collected in 2006–2007 from a PhD thesis, which examined the concept of innovative potential in the Hungarian Agricultural Biotechnology sector. The aim here is to understand how research organisations adapt and cope by examining how their activities have changed in recent times, in response to complications in the general and regulatory environment.

The chapter will initially look at the various complications that effect the innovation system. The data demonstrates three main complications; Section 8.2.1

F. Huzair (✉)
Development Policy and Practice Group, ESRC Innogen Centre, The Open University, Milton Keynes, MK7 6AA, UK
e-mail: f.huzair@open.ac.uk

P.T. Robbins, F. Huzair (eds.), *Exploring Central and Eastern Europe's Biotechnology Landscape*, The International Library of Ethics, Law and Technology 9,
DOI 10.1007/978-90-481-9784-2_8,

examines the situation of perceived lack in public demand for genetically modified (GM) crops and the expectations of the scientists who are the primary innovators in the system. Section 8.2.2 details the politics surrounding national science and technology policy and the fall in funding suffered by most of the publicly funded research institutes. Section 8.2.3 lastly discusses the politics behind national GM regulation and how uncertainty results from contestation, partly between Hungary and the EC, but primarily between interested parties within the country.

The chapter then moves on in Section 8.3 to look at the future as a source of uncertainty in the context and the problems this poses for long term planning and learning. Following this, Section 8.4 pulls from the data the strategies employed by various actors to cope with the difficulties outlined in Section 8.2 and an uncertain future as described in Section 8.3. The section finishes with a summary table showing the probable long term effects on the innovation system.

8.2 Environment Complexity and Uncertainties

Agricultural biotechnology innovation in Hungary occurs in a highly complex environment. There are overlapping environments that frame the space in which the network operates. European regulation creates a particular regulatory environment, as does the national regulatory environment within which it sits. There is also the market environment determined by domestic and international consumers who will use the products of the innovation system. Consumers in the innovation system are not only households, but also other actors, for example universities who may use patents that are discovered within Hungary or perhaps multi-national corporations. The position of each actor in relation to each environment will also vary. For example, if a research group is hosted within a university department or if they receive funding from European sources and not the national funding system, they may be relatively insulated from the changes in the national environment. Or for example an actor may be more or less tied and exposed to the instability of the market. Multinational corporations will be directly effected by the actions of national consumers, whereas basic research departments would be less so. Sensitivity to market conditions is partly dependent on how far upstream or downstream actors are in terms of the innovation process.

These conditions are common to many countries, but more unique to Eastern Europe and Hungary are factors such as uncertainty in funding linked to economic instability and contested national politics on the subject of GM and more generally around science and technology policy. Exploring these conditions in more detail, data analysis reveals that sources of uncertainty can be discussed in terms of three main types; (1) the public and demand for GM crops, (2) national science and technology policy and the fall in funding, and (3) regulatory policy uncertainty.

8.2.1 The Public and Demand for GM Crops

A strong market signal to those in the innovation system are the reactions of users of real or potential innovations. The most obvious problem in this context is the level of public hostility towards GM products generally.

A Hungarian PhD student working on a project comparing media coverage and attitudes towards Genetically Modified Organisms (GMOs) in Hungary, France and England reports that Hungary has a less negative approach to the GMO debate although there is generally thought to be less of an active or comprehensive debate than in either of these two countries (Personal communication, September 6). There is a lack of public involvement in Hungary possibly due to the non existence of forums in which such topics might otherwise be discussed. The farm scale evaluations in UK that took place over 3 years (the results of which were published in 2003) for example, involved extensive public consultation exercises which pushed the debate forwards. The resources required to conduct local and national public consultation on this topic are scarce and so this has not been allocated a high priority. Neither have attempts been made to identify, inform or consult with the farming community who form a politically important group. A former employee of the Ministry of Agriculture in the department of biotechnology when asked by the researcher if there are any other people the ministry should have been talking to but were not, answered in the following way:

> Yes of course, the farmers. They missed this step too. This was my other suggestion, to make questionnaires. To go out to the farmers and ask. Because if they don't want to use it, there is no point in thinking about it, or at least to know their opinion, but nothing, they didn't contact the farmers.

A report published in 2005 by Research International Hoffman, a market research company includes a survey of the 100 largest agricultural production companies registered by Monsanto in Hungary.[1] The survey shows that only 24% thought they were well informed on the topic but could not give an opinion. Twenty-nine percent said that they were not well, but satisfactorily informed and 46% said they were not well informed. Forty-six percent of survey respondents said that the reason that growers would choose a GM crop would be because of expected reduced production costs. Thirty-one percent said that higher yields would be the motivation for growing GM produce. In general 72% (representing 300,000 ha) said that they would produce GM crops if they were allowed to. This survey of course does not account for the thousands of much smaller farmers who are more likely to sell to a local market. There is also the opinion that if other members of the EU were to allow GM crops, then Hungary remaining GM free, would provide some sort of niche market and competitive advantage when exporting abroad.

With the absence of certain voices and groups, the debate appears to have been polarized by the increasing and active presence of non-governmental organizations

[1] Document in Hungarian, translated by AKI policy researcher

(NGOs). The policy stance held by Hungary, to ban the commercial cultivation of all GMOs and to prevent the field testing of certain varieties, is said by some to reflect the strong influence of NGOs. The former agricultural ministry employee suggests that the ministerial committee on biotechnology has a 60–70% membership from NGOs and the media with the minority from universities or the business sector (this point is corroborated by several interviews). True demand and market signals may therefore be disguised and a predicted level of acceptance should GM crop cultivation be allowed, is not known.

Activity by downstream innovators is increasingly inhibited by national law, the more applied the science is and so we might expect a certain future of no activity. However, scientists in the applied field continue with the opinion that given the general opposing trends of other member states and other countries globally the competitive pressure being applied to Hungarian farmers is increasing and so an opportunity to improve yields or lower costs, will eventually be accepted:

> Unfortunately the awareness of society regarding green biotech is badly effected ... at least in the short run is very difficult to judge what will happen. I'm an optimist, nowadays there is some small light that maybe the political and social attitude about green technology will be different soon. Interview 2006.
>
> If the innovation is ready for application, if it is very useful for peasant, for growers, it will breakthrough without any policy and against the NGOs. Canola can be grown in Hungary but the yields are very low compared with the neighboring Austria. And if a genetically modified canola can deliver a doubling of yields, it will break through. Interview 2006.

Additional signals confuse the picture such as the growing concern at the EU level in being able to compete in a global market in the growing biofuels sector. As reported in European Biotechnology News in June 2007:

> How can we remain competitive by producing biofuels with just conventional crops? One hectare of conventional corn produces around 6,000 litres of bioethanol, but the same hectare growing GMOs would produce around 14,000 litres – and the progress of research and development lead us to think that this gap will even be more important in the next few years. (Thierry de l'Escaille, head of the European Landowners' Organisation)

In the early part of 2008, food prices increased across the world due to the combined effects of poor harvests and competition for land between food crops and biofuels. The data collection for this project which ended in 2007 did not capture the influence of this new concern. This problem though, it may be imagined, would likely be called upon by those in favour of GM technology for food crops.

Lack of a domestic market effects less directly the scientific community than the sellers and distributors of agri-biotech innovations. Multinational agri-biotech companies who in other parts of the world, are engaged in the development and dissemination of GM technology, are not leaving Hungary. Apparently they stay because they have significant investments in the pesticide and conventional seed production and distribution market. It is with some certainty we can guess they are observing (at a distance) the debate and developments with much interest. Katz and Kahn (1978, cited by Birnbaum 1984) note that high technology companies in the US have adaptation strategies that include invention, diversification, increasing the size of the board of directors and joining trade associations. MNCs in this context show the same adaptive actions. In particular Monsanto is a member of the Hungarian Biotechnology Association, a strategic move grouping the company with pharmaceuticals which they hope will ensure an easier path to acceptance (interview 2006).

Smaller biotech companies that would also be involved in the stage that brings an innovation to the market do not have the resources to diversify, reinvent or employ the adaptation strategies that multi-national corporations do. They are therefore much more vulnerable to market change and uncertainty. This seems to be reflected by the very low numbers and high turnover of small and medium enterprises operating in the plant biotechnology area (research observation notes).

8.2.2 The Politics of National Science and Technology Policy

It has become unfashionable amongst economists to talk about "economic crisis" in the former Soviet States. The hardship that accompanied the initial stages of transition, the hyperinflation, massive unemployment and so on, certainly indicated a crisis situation in every sense that cannot be compared to the much improved general economic climate in more recent times. However, the term "crisis" has begun to reappear amongst those especially interested in the long term future of the science and technology system in Hungary. Problems with the national budget deficit surfaced in September 2006. The Hungarian Prime Minister Ferenc Gyurcsany, had been secretly recorded by a member of his own party admitting that he lied to the people and the country was on the verge of an economic crisis. The proposed changes to the government never occurred and the same economic crisis is ever looming on the horizon. The budget deficit crisis is not a blip and interviews show that it was not unexpected, it is the continuation of an underlying problem that stems from the Soviet era that the present government, like its predecessors, has simply failed to solve. This adds great uncertainty to national funding programs.

In an effort to reduce the budget deficit, the government has been reducing public spending in recent years. The fall in available funding fuels the debate on university closure. The large number of universities and colleges struggle to finance their extensive range of programs including molecular biology courses, so leading to the gradual closing down of these expensive departments over time. It has long been thought by many academic staff that the government should take action and strategically close some of these institutions (interview 2006). Although there has been no definite action in this direction, there has been some forced mergers of universities in the recent past. In 2006 the government recalled the funds that were distributed to universities, promising to return these funds later that same year (interviews 2006).

The lack of funding in plant science and the limitations that graduates face in the job market after completion of their university courses is reflected in student numbers. More students enrol on courses for human and animal sciences perceiving the pay and prospects to be better. Of those who do graduate with a PhD in plant sciences, a large number, in fact the majority, leave Hungary for destinations such as Germany, the US and the UK where their futures are more certain. This is creating a widening "generation gap" in the plant science research sector. The previous scientific generation socialised under the communist regime have settled in Hungary and have no intention to leave. This is the generation between the ages of 40 and 60 who are now heads of departments, university vice rectors, research centre directors etc. They are a small network and often know each other personally. The network is likely to become smaller as some of this generation approach retirement. As each new PhD graduate intakes completes, very few stay in Hungary to create the next middle generation of researchers and teachers. There is a widening gap which presents many difficulties for the future including being able to transmit the learning and skills required for new trajectories in molecular biology and innovation in cutting edge research.

There have also been discussions on a possible reorganisation of the Hungarian Academy of Sciences. This would be a yet more dramatic attempt to downsize the funding in addition to the year-on-year cut backs that many Hungarian Academy of Sciences (HAS) units suffer. The land occupied by the Plant Protection Institute (PPI) in Budapest may be sold by the government to raise funds, forcing the department to move to the countryside (research observation notes, 2006).

The general reduction in national funding is shown to affect directly the future survival of research institutes but it also has the potential to indirectly alter trajectories or directions in science and this will be discussed later in the chapter. Looking more closely at how national policy attempts to influence directly the direction of research and innovation in plant science, we can examine how science funding strategies and science and technology policies are actually perceived.

Rafols (2006) quotes the NKTH[2] budget at €83.1 million over 2002–2005, this is an average of just over €27 million per year. In 2008 the total budget for OTKA[3] was €20 million.[4] It is difficult to compare these figures or make a judgement about whether basic or applied science is better funded in Hungary. Rafols suggests that in the natural sciences, the HAS conducts around half of Hungary's R&D. One may argue that applied R&D is also carried out by private firms. However, Bross et al. (1998) adds that the relationship between science and industry is weak and that firms do not carry out enough R&D.

According to the NKTH website, the overall budget for science is being reduced in real terms. Basic science is being reduced (grants managed by OTKA and distributed through the Academy), but support for applied science is increasing (the

[2]The National Office of Research and Technology, Translated from the Hungarian: Nemzeti Kutatasi es Technologiai Hivatal.

[3]National Scientific Research Fund. Translated from the Hungarian: Orszagos Tudomanyos Kutatasi Alapprogramok.

[4]http://www.otka.hu/?akt_menu=991&set_lang=991

funds distributed by NKTH).[5] However, the conditions under which organizations can apply for applied research grants are proving difficult to fulfill (interviews 2006). Opinions both outside of this research and within this research remain divided as to whether basic science or applied science should be supported and promoted. However the NKTH is attempting to improve the rate at which innovations reach the market by means of funding strategies aiming to promote the applied sciences. Despite changes in policy, there is still a perceived lack of direction or national strategy in the area of plant science as is exemplified by the following interview excerpts from Hungarian agri-biotech scientists in 2006 and 2007:

> The problem is that as I mentioned, science priorities in this applied area are not well defined at the moment, so I don't see the priorities. Personally I dont see that there is any scientific strategy for science policy in the country
>
> Researcher: *"what is your opinion of national policy?"*
>
> (Interviewee): *"my opinion – there is no national policy"*.

The funding strategies for applied science show certainly a push from government agencies to increase applied science, innovations to market and to improve relationships between the science community and industry (literature analysis). However in this grounded piece of research which focuses on a particular community, we find that despite this, there is still a perceived lack of direction. This may mean one of two things; that the message is not being communicated effectively to this core group of people or that this community requires yet more specific direction which is not provided by the government. For example: what kinds of applied plant biotechnology would be successful in gaining funding? We find that both of these are true.

The general uncertainty created by the perceived lack of direction in national science and innovation policy is exacerbated by the uncertainty scientists face when applying for funding through national funding schemes. Certainly with regards to the applied funding grant application process there is very little way of knowing what types of research will be funded or why some projects are selected (corroborated by various interviews).

As an example of the perceived lack of transparency and direction, presented below are two extracts from interviews conducted with Hungarian agri-biotech scientists in 2006:

> For fundamental research its absolutely fair, but for the applied research, the transparency of how they are evaluating nowadays, is not fair. We don't know how they do it. And there are very mysterious calls about different topics, and we don't know why some are getting funding and not the others.
>
> It depends on the grant. This basic research grant, OTKA, this is perfectly transparent. This is the grant application we're working on right now. We didn't get the opinion of the reviewers and the only sentence they told us "this proposal does not belong to a sub task". And this is not true. ... This is only an excuse, we don't know what to change on the proposal. We tried to get more information, but failed.

[5]http://www.nkth.gov.hu/aktualis-hirek-esemenyek/kapcsolodo-cikkek/eastern-europe-struggles-080519

It may be expected that a lack of equipment would be the first indicator of economic struggles in the publicly funded research sector. This research finds that this isn't always true. Observations make apparent the concentration of research resources in fewer centres such as in Szeged, Martonvasar and the Agricultural Biotechnology Centre (ABC) in Godollo, but these laboratories are well equipped and in good condition. There are strategic collaborations both nationally and internationally which means that a lack of equipment rarely limits the activities of researchers. There are however some complicated policies which create difficulties:

> Sometimes they tell us that we can apply for new instruments and they give the sum minimum. You cannot apply for anything if it is below this minimum amount. We can apply just for huge instruments and we have to participate in [contributing] 30%. But from where can we get this 30% for such huge instruments? And this is every year we can apply and we are supported by the state, but nobody gives us this 30% so finally we can apply neither for the huge instruments or for the small instruments. (Interview Hungarian Scientist 2206)

Economic cutbacks and reductions in funding reverberate in the private sector. The university spin-offs, small and medium enterprises and public-private collaborations that might be part of a vibrant research intensive innovation system simply don't exist. In some cases universities do not have the resources to invest in technology transfer departments and in other cases with a shrinking research sector, there is not the critical mass to allow spin offs to flourish. University spinoffs must form with either some service or product and customer base in mind. In the agricultural biotechnology research sector these are limited due to market uncertainty or lack of information about the market. This is only one contributing factor explaining the lack of this type of activity, another notable factor is the lack of entrepreneurial culture (interviews 2006 2007).

8.2.3 The Politics of GM

In Hungary regulatory policy affects the private sector in a much more direct way than either a lack of domestic market demand or the economic cut backs. The Gene Law has effectively prevented the commercial cultivation of any agricultural GMO and in addition, the government has also protested against the field testing of certain varieties such as MON810. The case was referred to the European environment council and the decision was upheld. Although it was an interpretation of the precautionary principle that allows national law to be formed in this way, I argue in this chapter that the root of uncertainty is less the cause of ambiguity at the European level, but more due to instability at the national level.

Seemingly decided, the stance taken by the national government should at least provide certainty in that Hungary does not, and will not allow GMOs to be part of its future. This would allow scientists faced with this certain outcome to employ their skills in other areas and begin retraining for an alternative future. But national regulation is perceived as being incoherent and contestable. The nature of the decision making process itself throws a shadow on its ability to be stable over time.

In contrast to European regulation which is perceived as being "science based" and therefore stable, Hungarian regulatory processes and legislation are seen to be much more political in nature. As discussed above, regulation is subject to the influence of NGOs, farmers unions, and other politically motivated groups.

At the party level there are different possible approaches to the future of GMOs and GMO policy formation:

> Researcher: "*Are NGOs successful in influencing politicians?*"
> Interviewee: "*It depends on what the actual government is like. The conservative party is quite open to them and offered the Minister of Agriculture's chair for the leader of this collaboration. But this party has lost the elections. The socialist party, the leading power in the present government, doesn't put such an emphasis on them, they're quite apart from each other. One of the reasons for this distance keeping is their different views on Hungarian agriculture. This farmers' collaboration would like to see a change in agriculture so that the emphasis is being put on small or family farms and companies while the socialist party supports concentrated firms with mass production in agriculture.*"

The policy making process shows political in-fighting at the ministerial and committee levels. The competent authority for GMOs in Hungary is the Ministry for Agriculture. The Ministry for the Environment has the power to veto any decision made, a power that it exercises fairly frequently (interview 2006).

> Again the Minister of Agriculture wanted to make a kind of gesture to the opposition and created a 17 member group on co-existence. And out of those 17, 15 were organic growers, green organisations and two were the biotech association representatives. ... At the time he worked for the Godollo biotech centre. Professor ... stood up at the first session because it was so hostile. ... So Professor ... decided to leave the group right away and the result came in the form of the most restrictive policy stance proposal – a 400m isolation distance. (Interview 2006)

There is also reported what might be called the politicisation of the science itself. Scientific evidence selected for the support of certain regulatory proposals is accused of being questionable and gained through less than an unbiased independent process of scientific investigation. There are reports that individuals or organisations which generate scientific results are prone to political pressure or the promise of funding in exchange for producing evidence supporting either one side or another (personal communication, September 2006).

Hungarian scientists criticise the national regulatory framework for not being sufficiently science based. They compare the national regulatory framework to that of the EU, praising the latter for what they see as a science-based regulatory system. However, Hungarian policy researchers say there is a lack of "sub-politics" in the area of national GM regulation. In other words, the engagement of the public and the debate and discussion of policy at regional and local levels. This may be a reflection of the top-down nature of policy making and in particular, regulatory policy making. Policy makers at the ministerial level placate the influential national groups and NGOs who are able to bring a great deal of media attention, in an effort to maintain votes but simultaneously demonise the EU. Regulatory policy appears to be a tool or political space that allows the government to control various groups with an eventual aim to win favour with the voting public. The development of sub-politics, public debate, etc might weaken such control.

Jehlicka and Tickle (2004) predicted that CEE countries will become passively compliant with EU regulatory and governance requirements and their national perspectives will become eclipsed by EU hegemony. However, it is not the case that Hungary has simply chosen to passively accept EU direction in the area of agri-biotech regulation. Hungary has adopted a highly stringent form of EC regulation to meet its own ends, effectively and unexpectedly banning the commercialization of any GM crop in the country. Accession and the adoption of the Acquis Communautaire have not given national agri-biotech policy the predictability that might have been expected.

Regulation has differential impacts on innovation. For example in the energy industry, regulation often restricts the innovations that are possible but in the field of drug discovery, innovations are fewer where there is inadequate regulation and protection of intellectual property rights and patents. In Hungary in the field of agri-biotech it is less the regulation itself which is damaging to innovation, in some sense a narrow or strict regulation provides at least some direction to the path of innovation. But more damaging is the situation of the de-facto moratorium which creates a climate of uncertainty where those responsible for investment in a technology or expertise become hesitant to commit to what would be a sunk cost. This has the effect that only certain actors remain in the innovation system – large multinational firms that can afford the sunk costs of investment and who can apply the findings of research from and in other countries, and public sector researchers who have the choice, at least in the short term of spending allocated funding in non-applied areas of agri-biotech (see later in the chapter). Smaller private firms or research groups are no longer able to stay in the innovation system. It is possible to theorise that because smaller firms are more closely connected with the domestic market, are more sensitive to uncertainty at the national level.

8.3 The Future and Uncertainty as a Contextual Factor

One of the central tenets of national innovation systems theory (Edquist and Johnson 1997, Nelson and Rosenburg 1993) is that the national context and the country's specific history, economy and regulation shape the innovation system and its institutions. From the above discussion of the data, one specific factor of the national context emerges as being of significance. This is the relative uncertainty caused by the politics surrounding science and technology policy and policy on GM crops. Uncertainty relates to the future potential and use of an innovation in this field.

Rogers (1995) proposes that "technology is a means of uncertainty reduction that is made possible by information about the cause-effect relationships on which the technology is based". From a starting point of identifying such cause-effect relationships we assume that an innovation to create a GM crop is a solution to a given problem. These problems may be for example the need to increase yields, to improve pest resistance, drought resistance, to improve colour, texture or shelf life of a food.

i.e. there is a cause or demand, which has an effect of the development of a solution. In this context, there is no clear cause or problem acknowledged by policy. Data shows that individual scientists act on what appears to be altruistic motives, or developing solutions to problems identified by others amongst their acquaintance (farmers or plant breeding centres who tell them of a disease), or they are guided by a trajectory of expertise in a certain area. The problems which they might identify are not widely agreed upon and do not exist in policy. There is a degree of uncertainty which therefore cannot be reduced since the causes are not agreed and innovation cannot be expected.

There are however a certain set of problems that policy at least at the EU level has demonstrated are necessary to solve. Detection, bio-safety, traceability, and labelling are relevant examples. Therefore, innovation in this direction can be expected and indeed, the success of the bio-safety centre at Martonvasar is certainly evidence of this occurring.

Time is an important concept in the study of innovation. Technology development with regards to the influence of path dependence has been discussed to some extent in the literature. For example Nelson and Winter state: "the condition of the industry in each time bears the seeds of its condition in the following period." (Nelson and Winter 1982 cited by Antonelli 1997). In this section I also draw attention to the role of future expectations as a source of uncertainty in the shaping of a trajectory.

The activities of innovators are in part determined by expectations of the future. The relationship is similar to a demand-pull dynamic in that relationships exist between innovators and expectations about the future environment. Activities, routines, accumulated knowledge, patterns of learning, are all adjusted according to expectation of not only the future market and demands, but also the demands of increased or altered regulation, expected changes in alternative, inclusive or complimentary technologies and so on.

In the field of agricultural biotechnology there are many examples of this. The trend towards molecularisation means that research institutions have to engage in learning activities. Activities that assist in the learning process include attending conferences, taking visiting research fellowship positions and engaging in collaborative research. Research institutes such as the Plant Protection Institute also carefully consider the balance of young and older researchers. This is a somewhat expected change based on a predictable long term trend. The changes in market demand and regulation which are effected by politics, are much more difficult to gage. Uncertainty becomes problematic and planning becomes much more short term (research observation notes).

In the long term, uncertainty and the lack of long term planning can have serious consequences for the future of a research organisation. If learning and knowledge accumulation is not accomplished, innovative potential will be damaged and the institute can fail to re-engage in the technological trajectory as it moves forward, leaving it stranded with outdated physical and human resources.

The research institutions examined in this study have developed various strategies in order to cope with uncertainty and environmental complexity. As the next section will show, the strategy developed to cope with the various challenges of the context will have different consequences, so enabling the various complexities described above to be classified as more or less inhibitory to innovative potential.

8.4 Coping Strategies in the Hungarian Context

The agri-biotech innovation system and the various actors of which it comprises, in an environment without restriction, would face a vast sea of opportunities, directions and possibilities. Uncertainty poses a restriction. The evidence above suggests that any of these uncertainty created restrictions are not and in some cases, are not perceived to be, permanent. It is possible to visualise particular directions and areas in this spectrum of innovation possibilities as 'frozen'.

Figure 8.1 is a representation of the innovative actors and how one choice or direction may be chosen amongst others. Actors who are able to undertake activity are pictured as being within the webbed pentagram in the centre of the figure. Around them lies an entire spectrum of possibilities. Examples are given in the dashed ovals. However, due to national policy or otherwise, some of these possibilities maybe temporarily or permanently 'frozen' (indicated by *) so that actors are unable to access this research space. They may however redirect their attentions to alternative spaces.

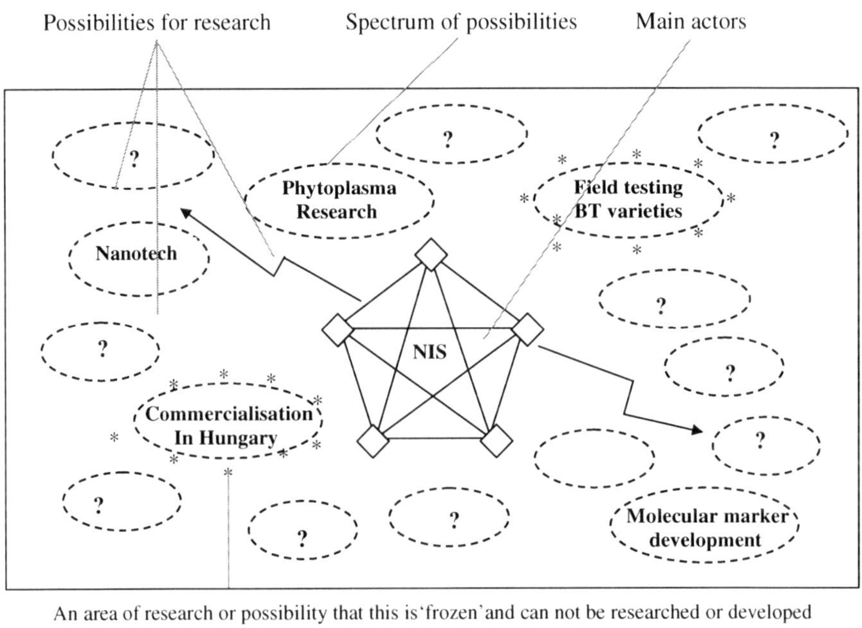

Fig. 8.1 Innovation possibilities and paths of actors

The effects on the actors are differential. There are three main strategies. (1) Actors may be unable to change their behaviour, possibly for reasons such as technological lock-in or a high degree of specialisation and so will either leave the system or look for markets outside the system. (2) Alternatively actors will adapt to the new conditions by changing their activities. And so the actors can either continue doing the same (and so they don't survive), continue doing the same, but outside Hungary, or (3) do something different. The more detailed descriptions of strategies given in the rest of this chapter are subdivisions of this very general overall description.

The data presents us with various examples of actor behaviour under the conditions of the complex Hungarian environment. Examining some cases in greater detail offers insights as to what motivates change and what allows adaptation.

Professor Ervin Balazs was a former director of the Agricultural Biotechnology Centre in Godollo. He began work in genetic engineering in the early 80s and has worked in France, India and South Africa. His realisation that Hungary at the time had very few protein and biochemistry engineers, which was the path the ABC was taking, led to the founding of a new research institute at Godollo in 2000. The aim was to develop a new line in environmental bio-safety research. The activities of the institute started with three divisions. The first was based on Ervin Balazs's own research interests, the development of new methods for detecting myco-toxins in animal feed and food. Second, a division to investigate whether food substances contain a GMO product, and a third division to investigate issues such as virus resistance and virus recombination. When asked what prompted this change in direction Balazs states:

> Because we had a very strong feeling that with this new technology, the public and society is quite eager to know whether this new technology has any impact on the environment and on human health.

Founding a new centre was based on two guiding instincts. One, that existing human resources would eventually limit the scientific trajectory that was carrying forward the ABC. And two, a sense that public demand would result in a new service market in bio-safety to run in parallel to the development of new technology. Such planning and development requires considerable foresight and knowledge of the sector. What may assist is an awareness of trajectories and trends in other countries. The development of a new division was also based on the accumulated knowledge and expertise gained in a specific field: developing methods for detecting myco-toxins. Success depended on being able to apply this skill to a new bio-safety market demand. Fundamental to such planning are key people who are experienced in the field and who have international connections. Also crucial is the ability to apply a foundation of knowledge in a new direction.

Such key figures illustrate the importance of the individual in the identification of potential pathways and dead ends in the formation of a trajectory (Nelson 2007). Although these actors take on some of the functions of a network organiser, these individuals are not sufficiently linked to other organisations and actors such as would make them effective network organisers as defined by Radosevic (1999). They are concentrated in the science system and are relatively upstream.

Ervin Balazs has since moved from Godollo to Martonvasar, the largest plant breeding institute in Hungary. He points out that in common with other research institutes there are two further trends occurring at Martonvasar. In response to the reduced funding, there is a move towards fundamental research rather than applied research. To conduct "fundamental" or basic research is much less expensive, it has been suggested that it may be eight to ten times so. The ABC in Godollo although originally set up to work in applied science, similarly shows some movement back towards basic research, though for different reasons. The ABC has had for some time, a programme of research in potato. By transferring a yeast gene into the potato they were able to achieve a good degree of drought tolerance. Due to national regulation which prevents the commercialisation of GMOs, scientists at the ABC have changed their research aims and outcomes. The aim is no longer to produce a drought tolerant variety, but they are using the variety they have produced to study the basic mechanisms of drought tolerance. They do this by comparing the drought tolerance they have created in their transgenic variety with the drought tolerance found in naturally tolerant wild species.

The ABC chose to launch this line of investigation and cope with the prevention of variety production in this way because it had accumulated a substantial body of knowledge following from a fairly natural trajectory in scientific investigation. The first work done on potato at the ABC looked at tuberisation. From this, the team began to look at the factors that effect tuberisation and found that sugars are important. From sugars they progressed to sugar sensing and then to drought tolerance. They predict that the next step will be research based on potato biodiversity. Niosi (2002) demonstrates that human learning is one factor that creates path dependence, and the same dynamics are exemplified here. Organisations such as the ABC invest in the development of expertise and capabilities in this particular area which is a sunk cost of an intangible type.

This research on the basics of drought tolerance conducted by the ABC reflects the second trend outlined by Balazs: that the output of the innovation system is increasingly in the form of the production of new knowledge and understandings via journal articles, papers and so on. Using the metaphor of the sand-clock:

> The difference between the American and the European sand-clock? When in the American sand-clock one grain of sand is dropping, it produces one dollar. In Europe it produces one paragraph. (Interview Ervin Balazs 2006)

And from the Plant Protection Institute in Budapest:

> The major output is papers. Scientific papers. But also with the breeding institutes, good collaboration, we cooperate to produce new resistant varieties, lines and so on. (Interview 2006)

This hints at a third type of activity taken forward by Hungarian research institutions: the production of new varieties via methods which are not perceived to be as harmful as genetic modification with foreign genes. The PPI would be working with breeding institutes to select on a genetic basis, lines which demonstrate particular required traits. The ABC in Godollo is simultaneously engaged in similar activities by including in their program of research, an investigation into drought tolerance as it occurs in wild species and is beginning to accumulate a body of knowledge in this field. The ABC is also using its expertise to produce molecular markers. This is a tool used by scientists that can be applied to assist the process of classical breeding. It allows varieties with required traits to be selected from very basic samples of plant material without having to grow the plants and select the varieties much further down the line through simple observation. Molecular markers therefore vastly improve the efficiency of selection without producing a genetically modified plant. Dr Janos Balint at Corvinus University is similarly engaged in what he terms "soft gene technology", the switching in or out of the plants own genes to alter function rather than the use of foreign DNA. He is explicit in the reasons for the direction:

> It is more acceptable for society or for green organisations... It's a necessity because the European Union don't want to accept gene technology in food science.

A fourth option is to continue to develop transgenic plant varieties but with the aim of producing for markets outside Hungary. Many large research institutes are involved with collaborative research projects with international partners to continue work with transgenic varieties. The work may not be completed within the country or only a component part of the overall project might be allocated to Hungarian partners. Such collaborations require scientists to have knowledge of global trends and to participate in networks (Huzair 2008).

A fifth new direction is gaining recognition. Actors are beginning to present GMOs as a solution to environmental problems. It is hoped that in this way public acceptance may be easier to achieve. Bioremediation is a growing market application for GMOs and a small number of Hungarian biotech firms are the first to see this as an opportunity. They seem to be particularly adept at marketing GMOs as a bioremediation solution to environmental problems and are ahead of the publicly funded research institutes in this respect.

Table 8.1 summarises the coping strategies employed and the activities undertaken by the various actors discussed above. The effect of these adaptations are also given.

Table 8.1 draws comparisons between the impacts on the innovation system of different types of uncertainty. Regulatory uncertainty and lack of internal market both have the effect of changing the activities of research institutes in particular. Rather than providing a limit to their activities, this type of uncertainty alters their trajectory, motivating them to search in areas outside the 'frozen' spaces of the innovation spectrum of possibilities. If alternative activities are related closely enough to the original activities undertaken in terms of skill usage and knowledge development, innovative potential is not lost, it is merely redeployed in other areas and there is the potential to re-engage in the original trajectory should it become a viable option in the future. However, this depends on how long the uncertainty will continue, as the increasing amount of time spent in alternative activities will create a permanent new trajectory as specialisms become established for example in bioremediation or molecular marker development. Similarly, returning to the practice of basic science and the publishing of papers sustains the possibility of re-engaging in the development of GMOs at some point in the future.

Table 8.1 Summary of coping strategies

Environmental Complexity	Main actors effected:	Is it short term or long term uncertainty?	Actor adaptation strategy	Effect of adaptation on the actors
Market Demand (the apparent rejection by the public, though the debate is highly polarised) And regulatory uncertainty	Down stream actors are highly effected by demand e.g. MNCs, biotech firms	It is a short term uncertainty, highly influenced by political trends at the national and international levels.	Actors perceive uncertainty as volatile, some publicly funded research institutes believe that GMOs will eventually be accepted and so do not change direction, but seek alternative global markets. They also develop alternative technologies e.g. molecular markers.	There is some change in trajectory towards alternatives for upstream actors – publishing papers, soft gene technology and complimentary technologies. Capabilities in the GMO area are continued and developed. Global networks developed. For down stream actors, particularly MNCs activity is completely prevented. Though since they are global players, they are not effected overall. Ties between upstream and downstream actors only temporarily disrupted – MNCs will not leave Hungary.

Table 8.1 (continued)

Environmental Complexity	Main actors effected:	Is it short term or long term uncertainty?	Actor adaptation strategy	Effect of adaptation on the actors
National policy (uncertainty A) funding and the economic crisis	Publicly funded research institutes, HAS and universities	A long term uncertainty resulting from the past and will continue in the foreseeable future	Larger research institutes that are globally connected can avoid some loss of funding through collaboration and EU funding Universities and unconnected research institutes shrink over time	The university system suffers permanent damage through the loss of departments, equipment and scientists who go abroad. Where there are no international connections, lack of funding can gradually cripple the innovation system and permanently damage innovative potential.
National policy (uncertainty B) Lack of direction in S&T policy	Research institutes and biotech firms	A short term uncertainty, related to economic development and government stability	The "wait and see approach" is applied. Preliminary plans for bioethanol and bioremediation have been made but cannot be implemented	The innovation system slows as the direction is unsure. Investment is limited. No guidance given with national funding and actors follow their own trajectories as far as they are able to.

The economic crisis and funding shortages have the most permanent and damaging effects on the innovation system. It is not the infrastructure or equipment that suffers, as collaboration offers a solution to this. It is the closure of university departments over time that will gradually produce fewer scientists and in addition to brain drain, will further increase the generation gap. Uncertainty in funding is paradoxically the most certain thing, as Hungarians are used to contextualising in the continuous debate over public spending and the economic crisis. As briefly mentioned above, the unexpected effect that funding uncertainty has had in this innovation system, is that it has forced the involvement of researchers and institutes in national and international research networks. Collaboration offers the opportunity for increased funding and Hungarian scientists with their stock of capabilities are successful in the development of collaborations and partnerships (Huzair 2008). This appears to be the only way that research institutions can circumvent the damaging effects of reduced government funding.

8.5 Conclusion

Lack of public demand for GM crops, uncertainty in funding and regulatory policy at the national rather than EU level, are factors which are causing adaptation and coping strategies to be employed that will shape the innovation system in agribiotech. The lack of funding clearly has damaging effects for a system so dependent on public funding. This is leading to the closing down of university biotechnology departments, the sale of government land, brain drain and the under-funding of key infrastructures such as university technology transfer departments. Over time this will lead to a shrinking of the innovation system as resources become more concentrated within the few well known large, state of the art facilities, such as the ABC in Godollo and the Bay Zoltan Foundation in Szeged. With the loss of smaller research groups, there is a loss of diversity in the areas of interest demonstrated by Hungarian agricultural biotechnologists. Any narrowing in the range of diversity, intuition tells us, is likely to result in a system that in the future will be less able to withstand significant change and shifts in demand.

The trend towards basic research, the publishing of scientific papers and so on, is one strategy that does not promote ties and connections within the actors of the system as well as would an applied research project. An applied research project would perhaps involve upstream actors who identify the practical problem, scientists and innovators who develop a solution, and downstream actors who would diffuse and distribute to a consumer or customer. The data viewed with this lens of coping strategies does not show a great deal of interdependence between actors, and in particular, down stream and up stream actors within the national system.

The system seems to favour the 'key individuals' as the drivers of institutional change. Nelson (2007) who states that innovation does not always follow a blind trajectory, but key individuals identify pathways, dead-ends and evaluate new technologies. This chapter provides evidence that this is happening in this study. While this is a good thing, the downside is that 'system learning' is not occurring to the same degree. Uncertainty, lack of funding and lack of direction in national regulatory policy does not allow for a stable environment in which such ties may be fostered with knowledge and innovations being developed across more than one organisation. In terms of the institutions that occur between the actors, what is seen is the isolation of organisations and the lack of co-evolution which has been argued, is important to innovation (Etzkowitz 1998, Etzkowitz and Leydesdorff 2000).

In an even more general way it might be noted that in the domain of basic research, within the directorship of key individuals or single organisations, the initial stages of investigation are carried forward by the interests and body of knowledge of the scientific team. It forms at first a fairly independent trajectory. However, as the science searches for an application, it becomes more exposed to legislation and is shaped increasingly by society and demand. I would suggest that it is more difficult to move from basic to applied research as has been attempted by large publicly funded, long existing organisations, than it would be for small private spinoff companies who are formed directly in response to a gap in the market and an unfulfilled demand.

The system as has been tested so far shows a remarkable degree of resilience. Evidence here suggests that by employing strategies of specialisation in fields such as soft gene technology, complimentary technologies e.g. biomarkers, bio-safety and so on, researchers so far have generally been able to engage themselves in activities which will allow them to re-enter the field of agri-biotechnology at some point in the future and preserve innovative potential, provided that the innovation system does not deteriorate past a crucial point and the structure of the innovation system (in terms of the 'middle generation'), maintains a sufficient integrity. Time matters not only because of uncertainty, but also because of irreversibility. Once key competences are lost, either through the loss of key individuals or otherwise, they will become difficult to replace.

References

Antonelli, C. 1997. The economics of path-dependence in industrial organisation. *International Journal of Industrial Organisation* 15:643–675.

Birnbaum, P.H. 1984. The Choice of strategic alternatives under increasing regulation in high technology companies. *Academy of Management Journal* 27(3):489–510.

Bross, U., A. Inzelt, and T. Reiss. 1998. Bio-technology audit in Hungary. *Technology, innovation and policy,* 221. Heidelburg: Physica-Verlag.

Chataway, J. 1999. Technology transfer and the restructuring of science and technology in Central and Eastern Europe. *Technovation* 19:355–364.

Edquist, C., and B. Johnson. 1997. Institutions and organizations in systems of innovation. In *Systems of innovation – technologies, institutions and organizations*, ed. C. Edquist, 41–60. London: Pinter.

Etzkowitz, H. 1998. The norms of entrepreneurial science: Cognitive effects of the new university-industry linkages. *Research policy* 27:823–833.

Etzkowitz, H., and L. Leydesdorff. 2000. The dynamics of innovation: From national systems and "mode 2" to a Triple Helix of university-industry-government relations. *Research Policy* 29:109–123.
Huzair, F. 2008. Innovative potential in Hungarian agricultural biotechnology and the evolution of networks. *Studies in Ethics, Law and Technology* 2(2):1–22.
Jeffries, I. 2002. *Eastern Europe at the turn of the twenty first century*. Routledge studies of societies in transition. London: Routledge.
Jehlicka, P., and A. Tickle. 2004. Environmental implications of eastern enlargement: The end of progressive EU policy? In *EU enlargement and the environment*, eds. J. Carmin and S.D. Vandeveer, 77–98. London: Routledge.
Nelson, K. 2007. Understanding economic growth as the central task of economic analysis. In *Perspectives on innovation*, eds.F. Malerba and S. Brusoni., 27–41. Cambridge: Cambridge University Press.
Nelson, R.R., and N. Rosenburg. 1993. Technical innovation and national systems. In *National innovation systems: A comparative analysis*, ed. R.R. Nelson. New York, NY: Oxford University Press.
Niosi, J. 2002. National systems of innovation are "X-Efficient" (and X-effective) why some are slow learners. *Research Policy* 31:291–302.
Radosevic, S. 1999. Transformation of science and technology systems into systems of innovation in Central and Eastern Europe: The emerging patterns and determinants. *Structural Change and Economic Dynamics* 10:277–320.
Rafols, I. 2006. *Inventory and analysis of national public policies that stimulate research in the life-sciences and biotechnology, its exploitation and commercialisation by industry in Europe in the period 2002–2005: National report of Hungary*. Brighton: University of Sussex.
Rogers, E.M. 1995. *Diffusion of innovation*, 5th edn. London: Freepress.
Von Tunzelman, N. 2005. Network realignment, policy sustainability and EU accession. In *Science and technology lessons for CEE countries*, eds. A. Inzelt, , G. Zaman, and S. Sandu. Bucharest: Expert Publishing House.

Chapter 9
Contested Agro-Technological Futures: The GMO and the Construction of European Space

Laurence Reynolds and Bronislaw Szerszynski

9.1 Introduction

In the opening years of twenty-first century, the European Union (EU) faced two challenges. On the one hand it was enlarging itself through admission of new countries to the Union, mainly in Central and Eastern Europe (CEE). On the other hand it was trying once again to harmonise its regulatory system for genetically modified crops and food, after an earlier attempt had been shattered by resistance from member states and civil society. In the early stages it was unclear how these two processes would interact with each other. Would accession countries such as Poland, Hungary and the Czech Republic be enthusiastic about the adoption of agricultural biotechnology? And, in the wake of the controversy, would the EU be able to resurrect itself as a homogeneous regulatory zone? Or would these two challenges combine to produce a more fundamental contestation over the future of farming and food in Europe? This chapter uses the way that this story unfolded to explore the relationship between models of agriculture and the regulation of agricultural technologies, using an approach which combines the sociological study of technologies and of space.

This chapter presents empirical data and concepts generated within the research project *Participatory Governance and Institutional Innovation* (PAGANINI), funded under the EU 6th Framework Programme for Research and Technology (Contract No. CIT2-CT-2004-505791), and also draws on work done as part of the research project *Facilitating Alternative Agro-Food Networks: Stakeholder Perspectives on Research Needs* (FAAN), funded under the EU 7th Framework Programme for Research and Technology. Earlier versions were presented to the conferences *The New Governance of Life: Challenges, Transformations, Innovations*, Vienna, 10–11 June 2007, *Regions and Regionalism in and beyond Europe*, Lancaster, 17–19 September 2007, and *The Promises and Challenges of the Life Sciences Industry in Central and Eastern Europe*, Prague, 18–19 October 2007. The authors are grateful to the other PAGANINI and FAAN project members, to participants at the above events, and to Les Levidow, Larry Busch, Piotr Stankiewicz, Andrew Barry and Bálint Balázs for helpful comments on earlier drafts.

L. Reynolds (✉)
Lancaster University, Lancaster, UK
e-mail: l.reynolds@lancaster.ac.uk

P.T. Robbins, F. Huzair (eds.), *Exploring Central and Eastern Europe's Biotechnology Landscape*, The International Library of Ethics, Law and Technology 9,
DOI 10.1007/978-90-481-9784-2_9, © Springer Science+Business Media B.V. 2011

Firstly, we analyse the GMO as implying a specific model of agriculture.[1] The conflict over different visions of agricultural futures in Europe has been theorised by Lang and Heasman (2004) as part of the 'food wars' where a dominant 'productionist' paradigm is being contested by a 'life-sciences integrated' paradigm and an 'ecologically integrated' paradigm. Marsden and Sonnino (2005) theorise this conflict in a similar way, but identify their paradigms slightly differently, starting with (i) a dominant 'agro-industrial paradigm' of intensively produced standardized agro-food commodities for global markets. This meets two newer competing trajectories – (ii) a 'post-productivist paradigm', where rural economies grow as leisure zones and (iii) a relocalised 'agrarian-based rural development paradigm' which draws on the sign values of quality production emphasising nature and locational specificity (see also Levidow 2008). We argue that GM strengthens the first paradigm, and further exacerbates its ecological contradictions by creating a continuous need for new pesticides – yet the way that the GMO is constructed as an abstract bio-legal entity tends to obscure its complex socio-ecological entanglements with models of agriculture and with wider ecological processes.

Secondly, we argue that the GMO also requires a homogenised regulatory space, encouraging the progressive abstraction of space and the erasure of difference between places: it favours what Manuel Castells (2000) calls the 'space of flows' over the 'space of place'. We trace the way that, as part of advancing the goal of enhancing the internal market, the EU's mechanism for regulating the movement of GMOs, the 1990 Deliberate Release Directive (DRD), attempted to create a standardised, 'striated' space of movement for the GMO within Europe.

We use this approach to tell the story of how the attempt to construct a standardised European space to match the standardised GM object met with resistance from member states and regions, as places resisted their further incorporation into the abstract space of the global agro-industrial model. We suggest that the DRD was further hampered by being shaped by both precautionary and innovatory imperatives, setting up a tension that would be played out in the ensuing GM conflict. In this conflict, the unified EU regulatory space began to break down, first with a series of national bans and then with an EU wide de facto moratorium. The EU then began a process of drafting new regulations around a new DRD which might allow for the lifting of the moratorium and the national bans. At the same time, the social contestation of GM had forged a powerful new alliance of environmental and consumer movements with important sections of the food retail and food manufacturing industries. This alliance was held together by the demand for the labelling of foodstuffs produced from or containing GMOs. We then go on to trace the way that the new round of regulation produced by these circumstances moved towards what we might call a 'regime of coexistence' whereby through labelling and traceability mechanisms the supply chains from 'farm to fork' would become segregated into GM and 'non-GM'.

[1] Although the term 'GMO' applies to any organism whose genetic material has been altered using recombinant DNA technology, in this chapter we are using it specifically to refer to genetically modified crops and food.

The previous cultural and scientific battles within Europe, especially around the demand for labelling, had strengthened the category of the 'non-GM', and allowed for a new route to added value for supermarkets, and a strengthening of alternative agrofood models around organic and local food networks. The battle would now be fought on the new terrain of 'coexistence,' with a new movement for 'GM-free regions' emerging. This movement proposed a strategy of quality agricultures, drawing on the sign values of 'clean and green' and of health, purity and the natural as a route to added value. Here, GM became constructed as a 'contaminant' that would threaten these alternative agro-economic strategies. Using the case of Poland in particular, we trace how the arrival of the new member states in Central and Eastern Europe intersected with these developments.

But first, in the next section we explain the approach we take to technology and space.

9.2 Technologies and Space

In this chapter we bring together two kinds of theoretical approach to our subject matter: we analyse how the GMO as a technology carries with it a certain model of agriculture, and we analyse how Europe has been constituted as an economic space of circulation through technological and regulatory mechanisms. Firstly, we draw on the sociology of science and technology to argue that GM crops should be analysed not simply as socially neutral technical means to the end of increasing crop yields, but as containing implicitly, within themselves, certain sets of social relations. Pinch and Bijker (1987) describe the 'social construction of technology' as a process of taking apart its 'black box' to uncover the social forces involved in its making. To supplement this view, Mackenzie and Wajcman (1999) argue against a formulation which one-sidedly emphasizes 'the influence of social relations upon artefacts' and therefore creates the opposite problem of a 'neglect of the valid aspects of technological determinism: the influence of technology on social relations'. To overcome this dualism, they declare that 'technology and society are mutually constitutive' (1999: 23). Thus technological developments reflect neither some 'logic of modernity' nor any other inevitable and exogenous force. Instead we must trace the hybrid social, economic, material, technical and cultural processes which construct, enact and maintain technologies in the world.

Following such an approach, technologies such as GM crops do not stand separately from society. Any new technology's creation is shaped by social relations – and in turn the new technology will shape social relations further. The main GM crops in circulation in the late 2000s were developed to fit a particular agricultural model; they carry the assumptions of this model with them into the world, and spread effectively only as part of that model. And at the same time – particularly because they are technologies based on living things – the material properties of GMOs mean that their interactions with the environment are highly complex and difficult to predict. However, in order that such complex objects could be rendered capable of movement around European space, they needed to be abstracted – turned into immutable mobiles (Latour and Wooglar 1986). The GMO had to be stripped of its cultural meanings and entanglements with ecologies and agricultures, and turned into a bio-legal entity capable of being identified and regulated across space and time (Lezaun 2006). The EU achieved this by developing a technocratic mode of regulation, which, as well as avoiding any WTO accusations of creating trade barriers, had the effect of 'black-boxing' the GMO – treating it as if its origins in very specific circumstances in the 1970s and 1980s, and the specific kind of socio-ecological relations that were assumed in its development, are irrelevant to its effect on the world.

Secondly, and relatedly, in order to understand the regulation of GM crops in the European Union we have to attend to the specific ways that European space is constituted. The governance of GM can be seen as an attempt to create a homogenous, abstract space fit for the circulation of similarly abstracted GMOs in a GM 'technological empire' (Barry 2001). In speaking of the 'abstraction' of European space we follow Henri Lefebvre (1991), who drew attention to the various modes in which spatial relations are experienced and organised within any given society. Lefebvre argues that space in the medieval period was experienced not as single, continuous and homogeneous, but as a complex discontinuous 'absolute' space fragmented by incommensurable bonds between people, places, symbols and divine forces. By contrast, the 'abstract' constitution of space characteristic of modernity is a space of equivalences and mobilities, a mathematically constituted, standardised space understood as a container for objects and activities, and one which privileges certain forms of knowledge and action, such as comparison, translation, exchange and movement. The modern state requires and makes possible this abstraction of space across a given territory, unifying it through a single system of relationships and duties, and a single logic of representation which simplifies reality across the territory and thereby makes it legible to state power (Scott 1998). The constitution of this abstract quality of modern space requires that social relations are progressively detached from their situatedness in given locales, a cultural achievement made possible through the proliferation of disembedding mechanisms such as money and expert systems (Giddens 1990).

An increasingly important mode and outcome of this abstraction since the late twentieth century has been the hypermobility of things. Manuel Castells argues that in the network society there is clash between two spatial logics – the space of flows and the space of place (Castells 2000: 407–459). The dominant spatial logic of network society is the space of flows and mobilities (Urry 2007), but this is always in tension with the space of places, of historically rooted human experiences (Tuan 1974). Jensen and Richardson argue that the dominant approach to European integration is one that favours that first spatial logic, by attempting to construct European space as a *monotopia* – 'an organised, ordered and totalised space of zero-friction and seamless logistic flows' (Jensen and Richardson 2004: 3). According to this analysis, the European project has become the endeavour to create a uniform space of flows of goods, services, capital, labour and technologies across the Union. The rise of this particular version of how to spatially cohere the European Union can be traced in the changing meaning of the term 'harmonisation' over the history of the EU. The earliest uses of the term used referred to the redistributive project of reducing social and economic differences between regions; however, as part of a general global trend toward neo-liberalism from the 1980s onwards, 'harmonisation' has increasingly been used to mean the removal of impediments to the international flow of objects, capital and labour (Barry 2001: 69–70). Ease of mobility across the European Union, rather than shared ideals or the redistribution of wealth, is coming to be seen as a key index of European cohesion; the GMO, in its regulation and diffusion, also follows this logic.

The two dynamics we have outlined – the abstraction of the GMO as a technology, and the abstraction of a space of circulation for that technology – occur together in a dialectical way, and one which depends on the establishment of 'machineries of equivalence'. For it would be wrong to conceive of modern, abstract space simply as an empty space of pure possibility that is progressively revealed when cultural, physical, political or economic barriers to movement and exchange are removed. To use the language of Deleuze and Guattari (1988), abstract space is not empty but *striated*, crossed and gridded by state power with vectors and metrics which prescribe functions to everything within it. Space that is abstracted in the way necessary for Jensen and Richardson's seamless flows to be possible is a dense space that is woven with connections and ligatures, a space created and sustained through complex machineries of equivalence and techno-regulatory mechanisms which allow these translations and mobilities to occur (see Levi-Faur and Jordana 2005). As David Harvey points out, for some things to be mobile, other things need to be fixed – whether a physical infrastructure (such as railways, ports and depots) or standardised administrative procedures such as laws, regulatory practices and treaties (Harvey 2001).

In the case of agricultural biotechnology, the harmonisation of definitions and regulatory practices, and the development of standardised machineries of detection and monitoring, has been an integral part of the project of constituting European space as a space of circulation for the GMO (Lezaun 2006). Later in the chapter, we will see how this project ran into difficulties; but in the next section we will see what the origins of today's GM crops in the 1980s tell us about the agricultural models that they imply.

9.3 A Glyphosate Planet: The Origins of Herbicide-Resistant GMOs

In the early years of the development of agricultural biotechnology, proponents of the new technology promised widespread and significant benefits, including giving plants the ability to fix nitrogen in the soil, thus not needing fertiliser, and creating plant varieties that were more resistant to drought and disease (OTA 1981). However, the GM crops that stand at the centre of worldwide controversy and contestation today consist overwhelmingly of plants with one or both of just two GM-added traits: plants modified to resist a proprietary herbicide, and plants given some insecticidal properties. Since the emergence of the technology onto the market in 1996, herbicide resistance has consistently been the dominant type of GM crop-trait used, and is deployed in 63% of all GM crops grown. Insect resistance has made up a further 18%, while plants with a combination of these traits make up most of the rest (ISAAA 2007). So, just as the global agricultural biotechnology sector is dominated by one corporation,[2] it is dominated by one kind of crop – herbicide-resistant varieties of bulk commodity crops such as maize and soy. And, as we shall see, the types of GM crop-traits that emerged have added to the dynamics of the controversy – raising questions of the possible intensification of the agro-industrial paradigm with its agrichemical lock-ins.

The GM crops that would form the focus of global controversy were developed in the 1980s within the labs of the 'university-industrial complex' (Kenney 1986) of North America and Western Europe. It was the Monsanto Corporation in the industrial heartlands of the USA that moved into the new technology earliest, most decisively and committing the most resources. Monsanto had stood at the heart of twentieth century American Fordism – occupying the key location of bulk commodity petro-chemical production. Yet by the late 1960s pressures were mounting that would force it to develop new technologies and markets. The GM crops it developed in the 1980s were its response to growing economic and ecological crises and contestations in the late 1960s and early 1970s, and as we shall outline later, this would shape the technology and the possible agro-social relations it bore along with it. Monsanto and its US rivals faced intensifying global competition from new low-cost bulk commodity chemical producers, prompting them to move up the technological hierarchy, increasing technoscientific innovation to capture the added value from knowledge-intensive production. This pressure had already prompted Monsanto's initial transition from production of bulk petro-chemicals into fine chemical production, especially pesticides (including herbicides) from the late 1950s onwards. Pesticides require high spending on R&D but also attract vastly greater prices per tonne (Heaton 1986: 236).

By the 1970s, Monsanto and other chemical corporations were dependent upon agrichemical production for an ever increasing proportion of their profits. In this period they had become an integral part of the transformation of agricultural production, which over the twentieth century had witnessed the gathering momentum of an 'agro-industrial model' originally developed in the USA (Goodman and Redclift 1991, Marsden 2003). This model is characterised by intensified production through increasing inputs from mechanisation, fossil fuels, fertilizers, agri-chemicals and crop-breeding sciences. These combine to produce large-scale specialised and homogenised cash-crop cereal and oilseed monocultures, intended for global commodity markets and intensive livestock production. To continue to occupy this location a company like Monsanto required a strategy of perpetual innovation. However, the development of new herbicides was becoming more difficult and

[2] In 2007 an estimated 87% of the world area planted with GM crops used Monsanto's seeds and traits (including Monsanto's GM technology licensed through other companies) (ETC Group 2008).

expensive by the 1970s, with all the simple and inexpensive chemical compounds having already been explored.

But Monsanto would be partly shielded from these economic problems by the discovery in 1970 of the herbicidal properties of glyphosate, which produced a broad spectrum herbicide. This was commercialised by Monsanto in 1974 as 'Roundup', a product that would become the world's biggest selling weedkiller, and one of Monsanto's most lucrative patented products, christened the first 'million dollar' pesticide (Heaton 1986: 256). This fulfilled a need for a perennial herbicide to fight the epidemic of wild grasses that, assisted by their vast rhizomal structures, had grown to fill the ecological niche left by the success of herbicides that controlled annual weeds and were causing a serious problem to farmers (Franz et al. 1997: 3). Glyphosate's systemic mode of activity proved to be highly effective against such perennial weeds. By 1982, 80% of the corporation's profits came from agri-chemicals, and their patent on Roundup earned Monsanto $500 million in that year alone. Roundup profits would largely underwrite the corporation's biotechnology research and development programme. As we shall soon see, these business dependencies around 'Roundup' would shape the development in the 1980s of the dominant GM crop trait, which would be glyphosate resistance, branded by Monsanto as 'Roundup Ready' GM crops.

The oil price rise of the early 1970s had hit Monsanto hard, cutting its operating profits by 88% (Collier 2000) and raising a desire within the industry to seek ways of moving beyond oil dependence. By the beginning of the 1980s Monsanto executives were moving to capture the promise of rDNA 'gene splicing' technology, announced by Boyer and Cohen in 1973, to address plant–pesticide interactions, one of its most important traditional areas of expertise. However, in order to achieve this Monsanto had to make a huge investment in basic research, and attempt to access, appropriate and adapt the latest scientific techniques and knowledge from university laboratories as it joined – and would eventually dominate – the race to produce the first GM plants.

But, having invested heavily in the basic science, by the mid-1980s Monsanto was keen to re-focus its efforts on finding an exit point, a key product that could begin to recoup this expensive outlay. The ability to find agronomically useful crop traits using rDNA techniques was limited, however. Most useful traits in crop plants are not reducible to a single 'gene', and may be the outcome of more complex interactions than even multigenetic explanations can account for (Fox Keller 2000). While such critiques of reductionism have been strengthened by recent developments in genomic sequencing and mapping, there have long been alternative views amongst scientists that in one form or another emphasise complexity. Thus in 1977 Pioneer Hi-Bred's director of plant research Don Duvick told a committee of the US National Research Council:

> Recombinant DNA molecular research probably will have little direct impact on the development of useful new crop varieties. Complex, delicately balanced interactions among many genes determine the phenotypes of successful crop varieties. [T]he in vitro recombinant DNA molecular techniques . . . are not suited to assorting and recombining very large numbers of genes into optimum genomic combinations (Duvick [1977] 2001).

Even Rob Fraley, the leading Monsanto scientist who developed GM crops, was later to admit that the utility of gene transfer technology was probably limited to simple traits such as resistance to herbicides and disease (Schell et al. 1989).

Of these two suggested traits, both are technically viable within the limited mono-genetic paradigm but one would also be commercially lucrative. Thus herbicide resistance emerged as the key trait to be aimed at. Glyphosate, being a broad spectrum herbicide, always had the obvious limitation that it killed both weeds and crops. Agri-chemical and seed companies had been attempting to develop seeds resistant to broad spectrum herbicides for some time, in order to approach plant–pesticide reactions from a different angle. Now rDNA plant biotechnology promised to confer this resistance, and develop a new plant–pesticide agricultural regime and business strategy.

At the end of the 1980s Monsanto finally found out how to use resistance developed by bacteria affected by pollution from around its Roundup production plants. Thus emerged 'Roundup Ready' herbicide-resistant GM crops, a technological choice shaped by a techno-social matrix with two main elements: the limits of the reductionist paradigm of monogenetic traits, and existing corporate techno-economic strategy around Roundup and plant–pesticide interactions. That is why out of all the promises of nitrogen fixation or drought and disease tolerance that animated early visions, this particular technological outcome triumphed.

As the new GM technology moved towards the market, Monsanto would increase its Roundup production capacity throughout the 1990s. In 2000, as its patent on the herbicide was about to expire, it would claim to have 'expanded its capacity to produce Roundup nearly five-fold since 1992' (Monsanto 2000) by investing in significant expansion of production capacity in its plants in Australia, Brazil, Belgium, North Carolina and Louisiana. In the last 5 years of the twentieth century sales grew by around 20% each year, by the turn of the century reaching about $2.6 billion annually, accounting for 67% of Monsanto's total sales (Panna 2002). In September 1998, Monsanto announced it was dropping the price of Roundup by between 16 and 22%, cutting $6–$10 of the price per gallon. At the same time it increased its technology fee for Roundup Ready soybeans from $5 per 50-pound bag to $6.50 (ICIS 1998). By boosting sales volume by lowering price Monsanto prepared to resist generic competition when Roundup came off patent, while indicating its new strategy of capturing value through the associated GM seeds.[3]

From this history of the production of the dominant GM crop trait of herbicide resistance, we can see that the new technology was shaped by a techno-social matrix that locked it into pre-existing economic and agro-industrial trajectories. The concept of a technological trajectory, as developed by Dosi (1982), enables us to trace how these different factors lock each other in to a fixed path. Chataway et al. (2004) find strong support for Dosi's concept in their work on understanding company strategies in agricultural biotechnology. They argue that:

[3]Hoechst would bring 'Liberty Link' to market in similar fashion.

> [e]ven in the context of what Monsanto managers perceived to be a radical change, there were strong links to technological and product strategies of the past.... Monsanto's radical technology and vision led strategy had its roots partly in its previous technology and product base (Chataway et al. 2004).

Recombinant DNA technology might have once promised a move beyond the ecological problems of the Fordist petro-chemical complex and its associated agro-industrial model. However, the new technology was shaped by the social forces around its creation into a component part of an intensification of the agro-industrial model. The GM crops that emerged from the corporate labs of America and Europe at the beginning of the 1990s implied a particular model of agriculture – an intensive monoculture maximising output for global markets and dependant on agrichemicals and other industrial and scientific inputs. The crops carry with them implicit agricultural practices such as intensive use of herbicides and other commoditised inputs. They have implications in terms of agricultural scale and the reduction of farm labour, reinforcing the growth of large-scale prairie-type agriculture and mechanisation. By enabling the bulk supply of cheap cereals and soy they reinforce an agrofood model characterised by intensive livestock and meat production, industrial food manufacture, long and complex supply chains, the persistent devalorisation of farm production, and continued 'cost-price squeeze' for farmers. This is reinforced by a technology agreement which legally obliges the farmer to engage in particular agricultural practices and banning the reuse of seed, thus locking farmers in to specific patterns of consumption.

The GMO thus emerged from the laboratories as a standardised object, around which the world would have to be remade. In contrast to alternative agricultural visions that draw on farmers' *art de la localité* to develop locally appropriate, sustainable forms of agriculture (Van der Ploeg 1993), the GMO exemplifies a 'scientific' approach to plant breeding which produces a standardised genotype that requires the simulation of a standardised agro-ecological environment around it, involving fertilisers, machines and pesticides. In the next sections, we will trace how the EU attempted to construct itself as a standardised space to allow the free circulation of this standardised object – and the problems encountered by such a project.

9.4 The Establishment of GM Regulation in the EU

In the last section we have seen how GMOs as agricultural technologies arose in a particular techno-economic context, and carry their origins with them. In this section we look at the European policy framework for the regulation of GMOs as it emerged in the late 1980s and has developed since then. Compared with regulatory practice in the USA, GM regulation in Europe has been seen as restrictive and precautionary – to the extent that it has provoked legal challenges through World Trade Organisation dispute settlement procedures, with the US in particular arguing that it constitutes an illegal barrier to free trade (Winickoff et al. 2005). But, as we shall see, EU GM regulation nevertheless had the effect of constituting European space in a way that is not neutral in terms of possible agricultural futures. In order to see this, we will first look briefly at the historical development of EU GM regulation, and the way it was shaped by two contrasting imperatives that operate within the European Union, in pursuit of environmental protection on the one hand and market-led technological innovation on the other.

In the late 1980s, the development of regulation of GM in the EU was uneven, with some member states such as Germany and Denmark advocating more restrictive regulation, and others such as Britain being more aligned to a form of trans-Atlantic neo-liberalism. But the EU's commitment to creating a single market within Europe led to the standardisation of GM regulation through a Deliberate Release Directive (DRD) (1990/220) which was more precautionary than countries like the UK would have favoured.[4] At this stage, the EU GM crop regulatory system thereby began to significantly diverge from that of the US, in particular by insisting that GMOs formed a distinct category requiring regulation. As part of the wider deregulatory move in the US at this time, in 1986 the OECD Blue Book had pronounced that there was no need for a special regulatory category of 'GM'. Yet by 1987 the European Parliament's Viehoff report was suggesting that there were 'special risks' attaching to GMOs because of their very origins in techniques of genetic modification. Thus, in contrast to the US regulatory framing which focused on them as *products*, to be regulated in the same way as equivalent products produced without rDNA technology, the dominant European approach became one that marked the GMOs as the products of a particular technoscientific *process*, and thus to be treated as a separate regulatory category (Jasanoff 2005: 45–54).

So by the time that the DRD came into force, it was framed in a precautionary manner – in effect, as an environmental protection directive. The significant 'Preamble' to the DRD states that:

> living organisms, whether released into the environment in large or small amounts for experimental purposes or as commercial products, may reproduce in the environment and cross national frontiers thereby affecting other member states; [and] the effects of such releases on the environment may be irreversible.

This framing led to a two stage procedure, involving experimental release prior to commercial release, covered by parts 'B' and 'C' of the directive respectively: Part 'B' covers experimental releases of GM crops such as field trials, while Part 'C' covers consent for commercial import, processing, feed or cultivation.

However, under the principle of free circulation of products within the internal market, and in order to lower the regulatory burden on the biotechnology corporations, the DRD also enshrined a policy that any 'Part C Consent' would be valid for the whole EU. For a Part C Consent, first a biotechnology company would submit a dossier of information (a 'Summary Notification Information Format', or SNIF) to the national competent authority of any particular Member State. Following a

[4]The precautionary character that the DRD took also shows the influence of the German presidency of the EU at a crucial time in its framing.

favourable opinion by this authority on the notification, the relevant Member State would then inform the European Commission on its opinion. If there are no objections raised by the other member states, the national competent authority that carried out the original evaluation then grants the consent. This consent, once given by the competent authority of any member state, would be valid for the whole EU.

Thus, the EU approach to regulating GMOs was shaped by a tension between two imperatives, those of precaution and innovation. Firstly, as the EU has become more of a state-like institution, it has progressively taken on the biopolitical functions of the modern state, those of protecting and optimising life within its borders (Foucault 2003). The environment has become a key framing for this imperative for the EU, not least because the trans-boundary nature of environmental issues has made them particularly suitable for the task of cohering the EU as a political body. Thus we can see the progressive incorporation of environmental protection and the precautionary principle in European Law, in the 1987 Single European Act (SEA) and the 1993 Treaty on European Union (TEU) or Maastricht Treaty. Although the precautionary principle is not mentioned by name in the 1990 DRD, the very fact that the Directive treats GM as a special, and risky, regulatory category, and advocates a two-stage process of 'deliberate release into the environment', underscores its precautionary logic.

Secondly, however, the EU functions in many ways as what Jessop calls a Schumpeterian competition state, a state that 'aims to secure economic growth within its borders . . . by promoting the economic and extra-economic conditions that are currently deemed vital for success', with a particular focus on 'technological change, innovation and enterprise' (Jessop 2002). This kind of state formation is particularly characteristic of the period since the 1970s, when the post-war accumulation regime of Atlantic Fordism – based on mass production, mass consumption, rising profits and rising incomes – went into crisis. The new Schumpeterian regime of accumulation, oriented towards competition, creative destruction and permanent innovation, was seen by capital as essential to recover profit rates and to stave off economic competition from newly industrialising countries.

Thus, in the EU, the *biopolitical* imperative to protect and optimise populations and ecologies is joined by an often-conflicting *technopolitical* imperative to optimise the far-from-equilibrium conditions seen as conducive to technological innovation. This latter imperative has had a powerful influence on the dominant economic vision for the EU, one which seeks to divert state support away from national and regional agriculture and manufacturing, and towards fostering the conditions for innovation. In more recent years, the Lisbon Agenda, and the more specific concept of the European Knowledge Based Bio-Economy (KBBE), itself a core part of the European Commission's 7th Framework Programme for Research and Technological Development (FP7), have further underscored the centrality of knowledge-intensive, high-technology industries to the dominant vision of the future prosperity of the Union. But such ideas had already been laid out in the Commission's 1993 *White Paper on Growth, Competitiveness and Employment*, and helped to shape the form of the DRD, in the way that the latter gave biotechnology companies a clear passage point at which to introduce a particular GMO into the whole of Europe. Even though the DRD marked out the GMO as requiring special regulation, once a specific GMO was deemed 'safe' by a member state, it would be free to circulate without specific post-release monitoring, labelling or traceability within the whole European Union.

Thus the attempt to create a standardised regulatory space for the global circulation of the GMO took a specific detour in the European Union. Although the OECD had advocated a wholly neo-liberal approach, the EU's need to find a harmonised regulatory framework which was consistent with the more precautionary instincts of member states such as Germany and Denmark forced it to make what would prove to be an unstable compromise between precaution and neo-liberalism – one which created a symbolic charge around the GMO as a form of 'living pollution', but did not follow through in terms of post-release monitoring and labelling. As we shall see, this internal tension within the DRD would work itself out in events across Europe.

But the DRD would also find itself in tension with socio-ecological complexity. The attempted constitution of a homogenous European space through and for the GMO operated through a number of interlinked modes of *abstraction*, using technical, administrative and economic mechanisms to try to allow the GMO to circulate without friction within Europe. This process involved the attempt to abstract the GMO from its social and ecological relations, and European space of its ecological and cultural diversity – and we shall see this put it in direct conflict with more local agricultural visions.

9.5 The Fragmentation of Regulatory Space

From the mid 1990s onwards the move towards EU consent for the first GM crop plants began to generate considerable controversy at many different levels: amongst member states, between member states and EU institutions, in the public sphere and in wider civil society. Amongst the concerns to be raised during the approval process by various member states and their scientific advisors were the possible environmental effects of the herbicide regimes related to the HR crops, the spread of GM plants as invasive species, the transfer of genetically modified traits to compatible species, the possible growth of resistance to the Bt insecticide amongst pests, harm to non-target insect species, toxicological effects and the spread of antibiotic resistance to bacteria. All these contestations fitted within the broad set of risk framings set out in the 1990 Deliberate Release Directive, around avoidance of harm to bodily health or ecosystems. This framing did not include important wider social effects on agricultural practices and rural social structure, including questions of scale and ownership and the model of agrofood system implied by the GMOs under consideration. Instead, the environmental and health framings had to bear the weight of social concerns around these wider issues.

These conflicts reached a crisis from 1996 onwards around the attempt to give consent to Monsanto's Roundup Ready soya and a herbicide-resistant and insecticidal (Bt) Maize owned by Ciba-Giegy (now Syngenta). These two varieties were the first crops to be commercially cultivated in the world, harvested in the USA in

1996. As they moved onto global markets and agro-ecosystems they began to galvanize open public controversy and intensify the pre-existing divisions within the EU. For our current purposes, the conflict around the (Ciba-Giegy/Syngenta) GM Maize variety illustrates the breakdown of the homogeneous EU regulatory space back into national units, while those around Monsanto's Roundup Ready soya illustrate the rise of the demand for labelling of GMO derived products.

The French government received the original notification for the GM Maize and passed on a favourable recommendation to the Commission. However when the Commission circulated this amongst the rest of the member states it received a large amount of objections from these states and their scientific advisory bodies. The 'Article 21' committee of representatives of the member states were unable to reach a majority opinion, so the Commission took the decision to the European Environment Council. When the Council met in June 1996, surrounded by the banners of protesting civil society groups, 13 out of the 15 member states, including even traditionally pro-GM governments like Britain, objected to the authorization. Spain was undecided – and according to a commission spokesperson even France, officially bound by its role as the proposing member state, 'was wavering'. However, without a unanimous decision the proposed crop variety could not be blocked and it was now within the Commission's powers to proceed with the consent, which it did in Jan 1997.

Despite this, the Commission's approval of the Ciba-Giegy/Syngenta GM Maize was seen to lack legitimacy amongst member states, leading to another important precedent. In the following months Austria, Luxembourg and Italy would invoke Article 16 of the DRD to place their own national ban on the maize. This move towards national bans in early 1997 would be repeated by many more member states around different GM varieties throughout the controversy. Also, following the Commission's decision to push through this first GM Maize variety in 1997, France decided not to authorise its cultivation domestically (though its sale was still permitted) and the European Parliament condemned the decision and demanded suspension of its import (Boy and de Cheveigné 2001). Thus even the overriding power of the EU commission to force a decision was not sufficient to gain widespread legitimacy or consent for this decision from many important member states, from the European Parliament or from wider civil society across the EU. The process therefore began to break down as member states imposed their own national bans, threatening to disintegrate any uniform European regulatory space.

The arrival of Monsanto's Roundup Ready soya marked an intensification of the controversy, drawing in the public and the media and escalating calls for the labelling of GM products. The first shipments of this GM soya were scheduled to arrive in European ports in November 1996. The demand for labelling was heightened by fact that the GM soya would be arriving mixed in with the conventional soya shipments (2% in the 1996 shipments, growing to 15% in 1997) and also by the centrality of soya or soya-derived ingredients for contemporary industrial food processing, which meant that the GM derived product would be spread amongst a huge array of food products. Furthermore, the USA was Europe's main supplier of soya, with the EU accounting for over 30% of US soya exports at the time.

Roundup Ready soya had already been given EU wide consent by the Commission for import as a food in May of 1996 in the face of opposition from Austria, Denmark and Sweden, who demanded labelling. Monsanto had submitted its 'notification' to the UK, as the UK was considered to have the most favourable government position towards GMOs in the EU (Charles 2001: 165). This body indeed made an initial favourable assessment in early 1995 and furthermore had argued there was no need for labelling. After its contested journey through the EU regulatory procedure, the Commission granted approval, also arguing that there were 'no safety reasons which justify the segregation of the product from other soya beans' and 'no safety reasons for labelling which mentions that the product has been obtained by genetic modification techniques'[5] – which put the EU in apparent harmony with the stance taken by the US administration, Monsanto and the American Soybean Association.

However, other actors within the food chain and wider society within Europe were taking a different view. In July of that year, EuroCommerce, representing a large section of the European retail, wholesale and international trade sectors, started to call for labelling on GM products, and other commercial organisations were to follow (FOEE 1996). In September of that year EuroCommerce held a press conference with the Greens in the European Parliament calling for a boycott of products made from the GM soybeans until these were adequately labelled. At the same time, environmental NGOs including Greenpeace began to mobilise public protest campaigns, and several large European supermarket chains and wholesale organisations declared they would not stock products containing the GM soya unless separated and labelled.

With rising European consumer, environmental and retail industry opposition to the GMOs, the EU nevertheless attempted to push ahead with the technology, repeatedly using its powers to overrule objections to new consents from member states, and sparking off more national bans on specific GMOs from Greece, France, Germany and Austria. While various national bans have ultimately ended up being condemned as without scientific basis by central EU bodies, the EU has been reluctant to take any sanction against member states, sensing it lacks the legitimacy around such a contentious issue. Thus five countries invoked Article 16 of the 1990/220 directive and even after this had been revised with the new 2001/18 Deliberate Release Directive, more national bans were enacted by Greece, Poland and Hungary.

The crisis intensified further in June 1999 when five EU member states – Denmark, France, Greece, Italy and Luxembourg – proposed a de-facto moratorium on any new Part C consents to the European Environment Council. The motion at Council said that, given concerns about risk, the specificity of European ecosystems, and the need to restore the confidence of public opinion and the market,

[5]Commission Decision 96/281/EC; Official Journal of the European Communities. 30.04.1996 – L 107 P. 0010 – 0011

the Commission should suspend new authorisations until it had strengthened and widened its risk-assessment procedures and put in place a system allowing the complete traceability of GMOs and products derived from them (CEU 1999). While the motion was not carried, this five-country blocking minority forced the Commission to make all further consents on its own, without the statutory backing of the Council, thus creating a serious problem of legitimacy for future consents.

In initiating what became a de facto EU-wide moratorium and series of national bans (and as the conditions for lifting these), EU member states called for a new regulatory framework around labelling, traceability and segregation of GM from non-GM food products. At the same time, the combined actions of retailers and social movements had helped to create a segregated market, with premiums for conventional non-GM and as well as organic and other 'quality' agrofood niches. These factors combined to produce a new attempted settlement – what we might call a 'regime of coexistence' within the EU, with the aim of creating parallel GM and non-GM agrofood chains from farming, storage and shipping through to manufacture, retail and consumption. This new regime would, it was hoped, restore the EU into a single harmonised bio-regulatory space for GM agrofood, overcoming its shattering into national member states.

Negotiations to end the moratorium culminated in a new version of the Deliberate Release Directive in 2001 (2001/18/EC). The revised Directive included changes in both principles of risk assessment (the consideration of wider and indirect effects; post-release monitoring; and restrictions on some anti-biotic resistance markers) and political mechanisms (the permissibility of ethical considerations; changes in the comitology; and more public consultation). These changes were supported by an explicit reference to the precautionary principle for the first time. Most significant for our story here is how the revised directive called for mandatory post-release monitoring requirements, with the preamble stating the necessity to 'establish common objectives for the monitoring of GMOs after their deliberate release or placing on the market as or in products', including 'monitoring of potential cumulative long-term effects'. These significant moves in the 2001 Directives towards labelling and traceability were further reinforced by the two additional regulations of 1829/2003 and 1830/2003.

As well as attempting to segregate GM and non-GM food manufacture and distribution, the same logic of segregation also began to pervade seed production and distribution, crop cultivation and the agricultural landscape. This led towards the adoption of a rhetoric of 'coexistence' within EU agriculture. Therefore in July 2003 the Commission published its guidelines on coexistence strategies for member states, which outlined ten general principles which were almost ostentatious in their embrace of a plurality of agricultures, signalling the partial dethronement of the previous dominant model of a singular, intensive, productivist, high technology agriculture for the EU. The first of these declared that: 'No form of agriculture, be it conventional, organic, or agriculture using GMOs, should be excluded in the European Union'. The second principle declared that: 'The ability to maintain different agricultural production systems is a prerequisite for providing a high degree of consumer choice', while the third clarified that: 'Co-existence refers to the ability of farmers to make a practical choice' between these three different forms of agriculture.

Unlike the earlier measures discussed, this was kept as a 'guideline', lacking the legislative force of an EU directive or regulation, a decision justified in the language of EU 'subsidiarity', with member states left to draw up their own national legislation (though the Commission made it clear that deviations would be met with legal sanctions). This move, in turn was justified in the guideline's appendix by reference to the diversity of agricultural practices and natural conditions across the European Union, making general rules impossible. Against this, oppositional NGOs and their allies in turn favoured an EU wide legislative framework on coexistence, with strict rules on liability. This exemplifies how within the apparent 'truce' of coexistence, the GMO battle rages on around questions of territory, scale and separation distances; between regional and farmscale segregation; around questions of thresholds for adventitious presence; and over whether the rationale for coexistence is an issue of physical or economic risk.

In the years following the publication of the coexistence guidelines member states began to devise their own frameworks, with the commission set to issue a further report on these in 2006. These negotiations formed a new battleground in different countries across the EU, around thresholds, separation distances and the question of scale – ranging from farmscale to regional and national proposed systems of agricultural segregation. Different member states would arrive at diverging proposed frameworks for coexistence, some following the commission's liberalism, others implicitly favouring non-GM quality regional agricultural strategies. Beyond this, regional governments started to become involved, with ten regional administrations approaching the European Commission in November 2003 with the demands that they could declare themselves 'GM-free Zones', and also able to enforce a polluter pays model of liability for contamination (Network of GMO Free Regions 2003). This network expanded to include 42 European regions by its 4th conference in 2005. It explicitly linked the agro-economic strategy of adding value via 'quality products' and certified regional identity to questions of preserving biodiversity and preventing GM contamination, first in its 2003 Brussels declaration and more substantially in its 'Florence Charter' agreed at another major gathering hosted by the Tuscan authority in early 2005 (Network of GMO Free Regions 2005).

Thus the question of coexistence began to act as a new focus for opposition to GM agriculture to spread across the EU, including many of the accession countries in Central and Eastern Europe. New alliances began to emerge, linking the regional government GMO-free grouping with dense networks of agricultural and civil society organisations. These groups came together for a founding conference in January 2005, announcing that over '100 regions and over 3500 subregional areas' from across Europe had made GM-free declarations (Haerlin 2005). Participants included ministers and officials from national, local and regional governments across the EU – many from important agricultural regional state administrations such as

Tuscany – and representatives of farmer and consumer networks and wildlife, conservation and environmental bodies (Levidow and Boschert 2008, Morgan et al. 2006). The conference issued a 'Berlin Manifesto' which spoke in terms of 'rights' for the 'regions of Europe' to determine their own ways of farming, eating, producing and selling food', and also of protecting their environment, heritage, seeds and economic futures.

In March 2006 the European Commission issued its report on the implementation of national measures on the co-existence of genetically modified crops with conventional and organic farming' (CEC 2006). Most of the member states had by that stage only begun preliminary draft versions of their coexistence rules. However, in a move that contradicted its earlier commitments to a subsidiarity-based approach to co-existence, the commission's report rejected 50% of this draft legislation proposed by Members States on the grounds that these would 'create obstacles to the free movement of goods'. The commission's 2006 report would prove controversial for a number of reasons. The document attacked the growing movement for GM-free regions, and hinted at legal action against member states and regions that tried to implement them. Furthermore, the commission favourably mentioned 'management measures that are applicable at the level of individual farms or in coordination between neighbouring farms' (CEC 2006). While the commission argued that GM-free zones discriminate against GM farming, critics noted how farm-level separation is much more likely to lead to GM contamination (FOEE 2006). In April 2006 Friends of the Earth Europe issued a detailed critique of the Commission's report, dubbing it a 'wait and contaminate approach' (FOEE 2006).

We have seen how by the end of the 1990s the combined actions of the 1990 DRD and social movement contestation had led to the unintended emergence of a segregated market, of GM and 'non-GM'. The new, post-2001 regulatory framework now attempted to institutionalise and stabilise this situation by creating parallel and separate agrofood systems – from cultivation, through shipping and processing, to retail – using a machinery of post release monitoring, labelling, segregation, traceability and coexistence. This move towards a 'regime of coexistence' exhibits important dynamics around questions of space, territory and scale. The initial blockage of GMO products in Europe was primarily enacted via the deployment of territorial tropes around the nation state. This is the meaning of the national bans invoked under 'Article 16' of the 1990 DRD (23 of 2001 DRD) where we witness powerful discourses of national sovereignty and defence of borders from alien pollutants. In its attempt to unblock the EU GMO moratorium and lift the national bans, the 'regime of coexistence' represents a shifting of the regulatory membrane. Now it moves away from national borders, and instead follows new contours that flow within and through nation states, between labelled products in grain silos and on supermarket shelves, and between GM and non-GM crops in fields. However, this new regime is still unsettled: As we have seen, the contestation over coexistence involved the rise of the GM-free regions movement, with the GMO struggle now once more respatialised and manifest as a battle over scale, between the largest possible territorial blocs proposed by GM opponents on the one hand, and the farm-scale attempt at molecular management favoured by the commission (see also Levidow and Boschert (2008) and Levidow and Carr 2010: chapter 8).

9.6 Accession: The Case of Poland

At the same time as the EU was attempting to roll out its new regime of GM governance based upon labelling and coexistence, it was also expanding eastwards. How would the accession of Central and Eastern European states affect the dynamics of these inter-EU contestations around coexistence? Furthermore, at this time the US government also launched its challenge to the EU's restrictions on GMOs at the World Trade Organisation (Winickoff et al. 2005). Would the CEE accession countries act within the EU in a manner more favourable to the US in this dispute? The dynamics of space, flow and locality would play out with particular intensity in some EU countries such as Hungary and Poland.

Polish agriculture has been a key battleground over GM and agricultural futures. In 2003 a team from the USDA (United States Department of Agriculture) visited Poland to assess its agricultural structure and potential on the eve of its accession into the EU (USDA 2003). They found that of all the accession countries Poland had the largest agricultural sector with 1.9 million farms, but that the majority of these were small family farms that averaged around just six to eight hectares each. While in the west of the country substantial areas of farmland were consolidated into plots of between 30 and 100 ha, 80% of Polish farms were below ten hectares. The USDA team contrasted Poland with the USA with its average farm size of 200 ha. Average farm sizes across the EU range from Greece with the smallest at 4.3 ha to the UK at around 69 ha.

The USDA found that most of Poland's farms were used to sustain families and local communities, rather than to produce for the market. Half of all farms were entirely subsistence operations while three-quarters provided only small or partial revenues for their owners, with most growing food to supplement income from full- or part-time jobs or pensions and around 25% of the population was involved in agricultural production. Based on this evidence the USDA team commented that:

> Poland's rural areas are over-populated. The rural population is highest in the southeast; farming there is very inefficient because of the high concentration of very small farms . . . To be competitive in the European Union, these smaller farms need to be consolidated into at least 30–40 hectare operations (USDA 2003).

However, the same USDA report also noticed the symbolic importance of the small, self-sufficient mixed farm in Poland and in central Europe generally. It suggested that 'Polish agriculture does not identify with large American-style farms specializing in growing just one or two crops'.

Apart from the size of the farms, the report remarked that 'Polish agriculture is not geared toward intensive commercial production. The majority of farms are mixed production entities growing fodder for their own animals'. Commodified agribusiness inputs were found to be low, with declining use of chemical fertilisers and many farmers saving and sharing seeds. The USDA report predicted that entry into the EU would bring exposure to globalizing agrofood trends and the world market that would lead to 'modernisation' and the overcoming of Polish 'backwardness' (USDA 2003).

In this linear conception of progress towards a single, globally universal agro-industrial paradigm, Polish agricultural employment and rural population would be reduced, farm sizes increased along with mechanisation and commodified agri-inputs, in order to produce cash crops for global markets. The USDA report predicted that the operation of the market would achieve these outcomes, a process accelerated by Polish accession into the EU. They add that 'farm managers will need to apply a market-oriented approach to agriculture, and to become more efficient' and that 'the European Union hopes small farms will naturally consolidate into larger operations from the market forces' (USDA 2003).

At the same time, social movements had begun to emerge around a desire to defend aspects of traditional Polish food, agriculture and rural social structure. In 2004 the International Coalition to Protect the Polish Countryside (ICPPC) launched an anti-GMO campaign. The ICPPC and other groups had already expressed worries over the negative effects of globalization and EU accession on Polish agriculture. The prospect of GM crops became a focus for these wider concerns. This new movement therefore began to lobby Polish local authorities to declare their regions 'GMO-free Zones', a move which was met with considerable success. By 2006 all 16 Polish voivodeships or provinces had made GM free declarations, putting Poland in the vanguard of the European GMO-Free Regions movement along with Tuscany and Austria (GRAIN 2008). Though without legal power, these declarations helped to crystallise and shape public opinion. In a parallel development, the Polish government adopted some of the strongest legislation in Europe, restricting trade in and cultivation of GMO seeds on Polish territory. This has subsequently become the focus of intense counter-lobbying by the US government and biotechnology industry bodies including Monsanto. It was also the subject of a challenge by the European Commission and the European Court of Justice, which proclaimed that the Polish legislation had no scientific justification. Poland has therefore become a key battleground for the future of GM crops in Europe, and implicitly over which agrofood models and whose social vision might flourish.

The pattern of rural social structure and farm size in Poland seems to play a role in shaping farmers' acceptance of or resistance to GM crops; but it is also having a more direct effect on the development of the regime of coexistence. Like the debates elsewhere in the EU, much of the discussion about Poland's GM laws has revolved around spatial issues; however, here the intensity has been if anything greater. In Polish debates around separation distances, zones and regions, the patchwork of family farms makes EU coexistence norms impossible. In 2008 the European Commission rejected a draft Polish law that would have restricted the planting of GMOs to designated zones. The Polish government had justified the measure by reference to the large number of small farms in Poland, arguing that this made the farm-level isolation of GM from conventional and organic crops impossible (GMO Compass 2008).

While the constellation of forces promoting GM agrofood consider Poland's family farms and rural agro-social structure to be a barrier to 'modernisation' and 'efficiency', other forces consider this instead to be an asset. In formulating their policies on GMOs, Polish legislators and regional authorities have emphasised the benefits of a GM-free 'brand' for trade and tourism. Even the USDA's 2003 report acknowledges Poland's potential as a leading European producer of organic foods, albeit within a heavily marketised framing (USDA 2003). Behind the debate about agricultural biotechnology and the GMO therefore we can see a wider debate that reveals a plurality of possible agrofood futures. Rather than a linear vision of progress towards the globalised industrial model implied by herbicide-resistant and pesticidal GM crops, other models and futures are also possible. One of these might be a strengthened organic sector, but within the framework of commodity production for the European market. However, beyond this, other agrofood patterns may become re-valued. While small family farms may not be 'efficient' at producing commodities for globalised markets, they may feed and sustain extended families and local communities in ways that do not register in simple economic statistics – especially in times of global economic turbulence and food crisis. Therefore family farms may add to social resilience and local food sovereignty. Furthermore, such a rural agro-social structure may have considerable ecological efficiencies and benefits when compared to the globalised and industrialised models.

9.7 Conclusion

There have been two interlinking themes in this chapter, both of which come together in the contested object of the GMO. One theme relates to space, place and mobility. The constitution of the GMO as a commodity involves the attempt to make it mobile, so that it can circulate according to a principle of free movement and free trade, along and across the striations laid down by 'regulatory capitalism'. But as an organism and a foodstuff the GMO is 'sticky' with biological and cultural connections to places – connected via gene flow, contamination, engaged publics and local agrofood cultures. As we saw, the global regulatory space constituted by the OECD became fractured by the very attempt by the EU to create its own harmonised regulatory market space out of uneven member-state regulatory postures with the 1990 DRD. However, for reasons explained above, European GM space swiftly fell back into the territorial logic of national and regional bans. A later attempt to re-harmonise EU space around labelling and coexistence by creating new regulatory membrane between farmers' fields and labels on supermarket shelves again ran up against the territorialised logic of GM free regions and local agrofood cultures, and generated a new battle over scale, involving farm sizes, separation distances and regions. This battle was intensified by the EU enlarging to include new member states such as Poland, often with very different patterns of farm structure and rural life.

The other theme relates to the relationships between technology, modernity, progress, and the possibility of alternative futures. We have argued that the herbicide-resistant GMO requires a world standardised to fit it, because of the industrial agrofood model that it assumes. But in the attempt to promote the GMO as the iconic technology of this model, we have seen how, ironically, other agrofood models have become valorised, and other futures made possible. As Marsden points out, a decade or more of intense contestation over GM has in fact strengthened local and regional embeddedness and provided greater coherence to alternative food and farming movements. Europe thus now enjoys a plurality of possible alternative agrofood futures, ranging from an organic market model, to local food networks. This story undermines the idea that there is one single linear path to one single technological future. The specific GMOs that we have were developed using early rDNA tech, and were always best suited for the post-war prairie agriculture characteristic of American states such as Iowa. Other kinds of knowledge, such as genomics, ecology and permaculture, are capable of suggesting very different technological options for future agriculture, and therefore very different worlds. Thinking about agricultural choices at the start of the twenty-first century is thus not just a question of discerning the one agro-technological future with which all societies must 'keep up,' but of asking which of a range of agrofood models people actually want, and which holds the most promise of enhancing the complex and differing functions that agriculture plays in any given places.

Bringing these two themes together shows that the struggle over agro-technological futures has a wider significance in relation to the tension that Castells describes in late modernity between the space of flows and the space of place. Castells argues that this tension threatens to produce a 'structural schizophrenia' between the networked, ahistorical space of flows on he one hand, and increasingly isolated and segmented places, increasingly disconnected from one another, on the other (Castells 2000: 459). This contrast seems to correspond to one between different agrofood paradigms – between the agro-industrial and the rural development paradigms described by Marsden and Sonnino (2005). And GM technology is far from neutral in relation to such contrasts, functioning as it does to stave off crises of accumulation suffered by the conventional agro-industrial model, and strengthening its tendencies towards concentration and transnationalisation, and thereby threatening more situated agricultures (Marsden 2008).

Yet the attempt to create an abstract space of flows and equivalences is always incomplete, and operates in a dialectic with situated forms of knowledge and activity. Delanty and Rumford (2005) argue for a more complex dialectic between flows and places than that implied by Castells, one in which there is a complex interpenetration of the logics of the global, the national and the local. Doreen Massey similarly argues that the overly stark contrasts between space and place such as those made by Tuan (1974) and Castells (2000) makes place too rooted, and thereby closed to reflexivity and to progressive social dynamics (Massey 2005: 183–184). The GM story in Europe confirms such arguments, suggesting that the dynamic between flows and places can produce unintended consequences that can expand the range of possible futures open to society. Counter-currents to the dominant agro-industrial paradigm are emerging across the post-industrial landscape of Europe, with demands for allotments, local produce and community-supported agriculture strengthening alternative agrofood networks and paradigms. Such developments may often be fragile in the face of established economic and political forces, but singly and cumulatively they show that, if the agro-industrial model triumphs, this will not be simply because there are no other options. From this standpoint, the rural agro-social structure of CEE nations such as Poland is not necessarily a pre-industrial hangover to be brushed aside, but might instead be a harbinger of possible post-industrial agrofood futures and paths towards alternative, ecological modernities.

References

Barry, Andrew. 2001. *Political machines: Governing a technological society*. London: Athlone Press.

Boy, Daniel, and Suzanne, de Cheveigné. 2001. Biotechnology: A menace to French food. In *Biotechnology 1996–2000: The years of controversy*, eds. George Gaskell and Martin W Bauer, 181–190. London: Science Museum.

Castells, Manuel. 2000. *The Rise of the network society*. Oxford: Blackwell.

CEC. 2006. *Report on the implementation of national measures on the coexistence of genetically modified crops with conventional and organic farming*. Brussels, 9.3.2006. COM(2006) 104 final. http://www.ec.europa.eu/agriculture/coexistence/com104_en.pdf

CEU. 1999. *Draft minutes of the 2194th Council Meeting (Environment)*. Council of the European Union. http://register.consilium.europa.eu/pdf/en/99/st09/st09433-re01.en99.pdf

Charles, Daniel. 2001. *Lords of the harvest: Biotech, big money, and the future of food*. Cambridge, MA: Perseus.

Chataway, Joanna, Joyce Tait, and David, Wield. 2004. *Understanding company R&D strategies in agro-biotechnology: Trajectories and blindspots*. Research Policy, 33/6–7, 1041–1057

Collier, Robert. 2000. Regulation of rbST in the US. *AgBioForum*, 3(2&3):156–163. http://www.agbioforum.org/v3n23/v3n23a14-collier.htm

Delanty, Gerard, and Chris Rumford. 2005. *Rethinking Europe: Social theory and the implications of Europeanization*. London: Routledge.

Deleuze, Gilles, and Félix Guattari. 1988. *A thousand plateaus: Capitalism and schizophrenia* (trans: Massumi, B.). London: Athlone Press.

Dosi, G. 1982. Technological paradigms and technological trajectories: A suggested interpretation of the determinants and directions of technical change. *Research Policy* 11(3):147–162, June 1982.

Duvick, Donald N. [1977] 2001. *Recombinant DNA molecule research potential for agricultural crop plants*. Testimony to National Research Council – April 11, 1977 – Transcribed by D. N. Duvick, March 7, 2001. Available online at: http://www.mindfully.org/GE/RDNA-Ag-Potential-Duvick.htm

ETC Group 2008. *Who owns nature? Corporate power and the final frontier in the commodification of life*. Ottawa: ETC Group.

FOEE (Friends of the Earth Europe). 1996. *Biotechnology programme mailout*, Vol. 2, Issue 6, 15th September 1996.

FOEE (Friends of the Earth Europe). 2006. *Contaminate or legislate? European Commission policy on 'coexistence'*. Position Paper April 2006. http://www.foeeurope.org/publications/2006/contaminate_or_legislate.pdf

Foucault, Michel. 2003. *Society Must Be Defended': Lectures at the Collège De France, 1975–76* (trans: Macey, D.). London: Allen Lane.

Franz, John E., Michael K. Mao, and James A. Sikorski. 1997. *Glyphosate: A unique global herbicide.*, ACS Monograph 189. Washington, DC: American Chemical Society.

GMO Compass. 2008. Poland may not ban genetically modified plants, *GMO Compass*, Jan 2008. http://www.gmo-compass.org/eng/news/325.poland_may_ban_genetically_modified_plants.html

Gaskell, George, and Martin Bauer. eds. 2001. *Biotechnology 1996–1999: The years of controversy*. London: Science Museum Press.

Giddens, Anthony. 1990. *The consequences of modernity*. Cambridge: Polity Press.

Goodman, David, and Michael Redclift. 1991. *Refashioning nature: Food, ecology and culture*. London: Routledge.

Grain (Genetic Resources Action International). 2008. 'Polish Farmers Defy EU Bureaucracy', *Seedling Magazine* July 2008. http://www.grain.org/seedling/?id=558

Haerlin, Benedikt. 2005. *European conference for GMO-free regions, biodiversity and rural development*. Press Release 24/01/2005, Berlin.

Harvey, David. 2001. *Spaces of capital: Towards a critical geography*. Edinburgh: Edinburgh University Press.

Heaton, CA. 1986. *The chemical industry*. Glasgow and London: Blackie and Son.

ICIS. 1998. 'Monsanto Cuts Its Price for Roundup' *ICIS News*, 07 September 1998. http://www.icis.com/Articles/1998/09/07/88691/monsanto-cuts-its-price-for-roundup-to-boost-demand-throughout-the.html

ISAAA. 2007. *Global Status of Commercialized Biotech/GM Crops: 2007*. http://www.isaaa.org/resources/publications/briefs/37/executivesummary/default.html

Jasanoff, Sheila. 2005. *Designs on nature: Science and democracy in Europe and the United States*. Princeton, NJ: Princeton University Press.

Jensen, Ole Brandt, and Tim Richardson. 2004. *Making European space: Mobility, power and territorial identity*. London: Routledge.

Jessop, Bob. 2002. *The future of the capitalist state*. Oxford: Polity Press.

Keller, Evelyn Fox. 2000. *The century of the gene*. Cambridge, MA: Harvard University Press.

Kenney, Martin. 1986. *Biotechnology: The university industrial complex*. New Haven, CT: Yale U.P.

Lang, Tim, and Michael Heasman. 2004. *Food wars: The battle for mouths, minds and markets*. London: Earthscan.

Latour, Bruno, and Steve Woolgar. 1986. *Laboratory life: The construction of scientific facts*, 2nd edn. Princeton, NJ: Princeton University Press.

Lefebvre, Henri. 1991. *The production of space* (trans: Nicholson-Smith, D.). Oxford: Blackwell.

Levi-Faur, David, and Jacint Jordana. 2005. Regulatory capitalism – special issue of. *The Annals of the American Academy of Political and Social Sciences* 598:1.

Levidow, Les. 2008. European quality agriculture as an alternative bio-economy. In *Reconstructing biotechnologies: Critical social analyses*, eds. Guido Ruivenkamp, Shuji Hisano, and Joost Jongerden, 185–205. Netherlands, Wageningen Academic. http://technology.open.ac.uk/cts/docs/LL_Quality%20Agriculture_08.pdf.

Levidow, Les, and Karin Boschert. 2008. Coexistence or contradiction? GM crops versus alternative agricultures in Europe. *Geoforum* 39(1):174–190, 2008.

Levidow, Les, and Susan Carr. 2010. *GM food on trial: Testing European democracy*. London: Routledge.

Lezaun, Javier. 2006. Creating a new object of government: Making genetically modified organisms traceable. *Social Studies of Science* 36(4):499–531.

MacKenzie Donald and Wajcman Judy. eds. 1999. *The social shaping of technology*. Second edition. Buckingham: Open University Press.

Marsden, Terry. 2003. *The condition of rural sustainability*. Assen: Van Gorcum.

Marsden, Terry. 2008. Agri-food contestations in rural space: GM in its regulatory context. *Geoforum* 39(1):191–203, 2008.

Marsden, Terry, and Sonnino Roberta. 2005. Rural development and agri-food governance in Europe: Tracing the Development of Alternatives. In *Agricultural governance: Globalization and the new politics of regulation*, eds. V. Higgins and G. Lawrence, 50–68. Abingdon: Routledge.

Massey, Doreen. 2005. *For space*. London: Sage.
Monsanto. 2000. *A Single Focus: 2000 Annual Report*. http://www.monsanto.com
Morgan, Kevin, Terry Marsden, and Jonathan Murdoch. 2006. *Worlds of food: Place, power and provenance in the food chain*. Oxford: Oxford University Press.
Network of GMO Free Regions. 2003. *Brussels Declaration*, November 2003. http://www.gmofree-euregions.net:8080/docs/ajax/ogm/Brussel_declaration_nov2003_EN.pdf
Network of GMO Free Regions. 2005. *Florence Charter*. http://www.gmofree-euregions.net:8080/docs/ajax/ogm/Charter_en.pdf
OTA. 1981. *Impacts of applied genetics: Micro-organisms, plants, and animals*, April 1981. US Congress Office of Technology Assessment. Washington, DC: US Government Printing Office.
PANNA. 2002. *Monsanto corporate fact sheet. Pesticide action network North America*. http://www.badseed.info/resources/monsanto.pdf
Pinch, Trevor, and Wiebe Bijker. 1987. The social construction of facts and artefacts: Or how the sociology of science and the sociology of technology might benefit each other. In *The social construction of technological systems: New directions in the sociology and history of technology*, eds. Wiebe Bijker, Thomas Hughes, and Trevor Pinch, 17–50. Cambridge, MA: MIT Press.
Schell, Jozef, Bruno Gronenborn, and Robert T. Fraley. 1989. Improving crop plants by the introduction of isolated genes. In *A revolution in biotechnology*, ed. Jean Marx. Cambridge: I.C.S.U/Cambridge University Press.
Scott, James C. 1998. *Seeing like a state: How certain schemes to improve the human condition have failed*. New Haven, CT: Yale University Press.
Tuan, Yi-Fu. 1974. *Topophilia: A study of environmental perception, attitudes, and values*. New York, NY: Colombia University Press.
Urry, John. 2007. *Mobilities*. Cambridge: Polity.
van der Ploeg, J. 1993. Potatoes and knowledge. In *An anthropological critique of development: The growth of ignorance*, ed. M Hobart. London: Routledge.
USDA. 2003. *Poland: Basic agriculture. past, present, and thoughts on its future in the European Union*. http://www.fas.usda.gov/pecad2/highlights/2003/12/Poland/index.htm
Winickoff, David, Sheila Jasanoff, Lawrence Busch, Robin Grove-White, and Brian Wynne. 2005. Adjudicating the GM food wars: Science, risk, and democracy in world trade law. *The Yale Journal of International Law* 30(1):82–123.

Chapter 10
Final Remarks

Peter T. Robbins and Farah Huzair

10.1 Introduction

The biotechnology landscape in Central and Eastern Europe is shaped by its communist history, and the ways individual states have responded to global market transition in the context of European Union accession. In Chapter 1 we described the influence of centralized control and planning of the Soviet regime on the development of science and technology in Eastern Europe. The R&D sector was large, though not innovative and research in the sciences was geared towards disciplines with militaristic applications, e.g. physics and chemistry. In one or two exceptions, biology and biotechnology expertise was developed and maintained (e.g. agricultural biotechnology in Hungary), though generally the life sciences suffered neglect and worse, the setbacks of Lysenkoism.

The soviet academy model, developed in Russia and rolled out across the Soviet satellite states, comprised a system whereby the academy undertook mostly basic research and government research institutions carried out the bulk of applied R&D. Catalogues of innovations and inventions would be passed to industry for selection and use. In most (but not all) countries, universities were reduced to mere teaching factories, playing no integral part in the R&D system. There was little in-house industrial R&D (with the exception of the Czech Republic and Slovakia (Radosevic and Auriol 1999) see Chapter 2 by J. Senker et al., this volume).

Transition to a market economy followed different models and principles in various former Eastern European states. Hungary for example took a gradual approach to transition with reforms beginning in the 1960s and more intense reforms occurring in the 1990s. It is argued that a gradual phasing out allows containment of small errors and corrective action as policies change. This approach should also allow a better understanding of the capitalist system to develop. Countries at the other end of the scale such as Poland, demonstrate a 'shock therapy' approach. On a far shorter time scale measures involve firstly liberalisation to include the freeing of prices,

P.T. Robbins (✉)
Development Policy and Practice Group, ESRC Innogen Centre, The Open University, Walton Hall, MK7 6AA, Milton Keynes, UK
e-mail: p.t.robbins@open.ac.uk

P.T. Robbins, F. Huzair (eds.), *Exploring Central and Eastern Europe's Biotechnology Landscape*, The International Library of Ethics, Law and Technology 9,
DOI 10.1007/978-90-481-9784-2_10,

foreign trade and the end of central planning. Second, a program of privatization of state enterprises, and third, measures to promote macroeconomic stabilisation. Advocates argue that rapid change provides unambiguous signals of intent in the direction of reform and the bypassing of bureaucracy with rapid reductions in state regulation helping to deter corruption. It is also possible that following a change in government, leaders are able to use a 'honeymoon period' shortly after election to make irreversible, comprehensive and often painful measures towards permanent reform. Unfortunately the idea of change can become associated with hardship (Jeffries 2002).

This volume has shown that discussion and policy development on the introduction and integration of the life-sciences in CEE societies is important in the context of EU integration and economic development. It has been framed as a key enabling technology, a driver of innovation (Chapter 2 by J. Senker et al., this volume) and an engine of growth (Lyall et al. 2009). To conduct discussions and engage society on life science policy is problematic for every government as they are faced with questions surrounding deliberation and governance; who should be involved, how they are to be represented, in what sort of processes etc. For CEE, a further issue to contend with is the continuing erosion of governance capabilities. In many former soviet states, the process of transition in the 1990s was accompanied by economic hardship due to the austerity measures put in place in line with IMF guidelines (Jeffries 2002). Public debates and other exercises in deliberative democracy and participative policy making are resource intensive. They require infrastructure and arguably, are more successful if there is existing experience and expertise. Not only are financial resources scarce in transition economies, so too is the experience of modern practices of govern*ance* as opposed to traditional forms of centrally planned govern*ment*. This problem is possibly exacerbated by the withdrawal of the state from many policy areas at the time of transition, in order to reverse the extremely interventionist role of the state in times prior. This may though, have resulted in a transitionary period where the state did not intervene enough (Von Tunzelman 2005).

Lyall et al. (2009), examining cases of governance of the life sciences elsewhere in the world, find that the challenge in governance relates to the important role of non government actors in the policy process and the evolving nature of state-society relationships in policy making which more closely resemble networks than hierarchies (Bache 2003). A study of the life science industry in CEE, its development, current state and future prospects, was begun in a 2008 special issue of Studies in Ethics Law and Technology.[1] Accounts focused on industry, innovative capabilities and the upstream inputs that lay the foundations for a successful life science sector. This volume is a continuation of that narrative. It is an examination of the evolving state-society relationships in CEE through an analysis of the roles of actors such as experts, the public, the media, government policy and law, and

[1] Special edition of Studies in Ethics, Law and Technology. Volume 2, Issue 2, August 2008 http://www.bepress.com/selt/vol2/iss2/

other non-governmental organisations like the church. This volume also explores the duality of governance frameworks alongside maintenance of a necessary conventional command and control style of regulation which is based in legislation, evidence-based decision making (Lyall et al. 2009) and expertise.

10.2 Economic History and the Evolution of Political Institutional Relationships

As we mentioned above, following the opening up of CEE economies, austerity measures in line with IMF guidelines were commonly introduced to combat hyper inflation. Measures included expenditure cuts, limits in public sector wage increases, currency devaluation and import duty increases (Jeffries 2002). A World Bank study of 5 post communist states (Poland, Romania, Russia, Hungary and Germany) completed in 1999, suggests that international support was driven by an implicit model of minimalist, "hands-off" public administration. Authors of this study describe a wave of anti-statism resulting from the delegitimation of the communist state and a "prevailing intellectual wind" from many developed economies at the time (Nunberg et al. 1999).

These trends, characteristic of many OECD countries in the 1980s, were described by Hood (1995) as a movement towards New Public Management (NPM). Some of the doctrines underpinning NPM include; disaggregating public organisations into separately managed corporatized units, greater use within the public sector of management practices sourced from the private sector, finding less costly ways to deliver public services, more "hands-on management" (that is, more active control of public organisations by visible top managers wielding discretionary power), and a move towards more explicit and measurable standards of performance. These are in line with international so-called "megatrends", for example; attempts to slow down or reverse government growth to reduce public spending and the shift toward privatization and away from core government institutions, with renewed emphasis on subsidiarity in service provision (Hood 1991).

On one hand, advocates of NPM cite the benefits of a politically neutral or apolitical system (Hood 1991). However, a commonly felt implication in developed economy administrative systems was a general unease or tension between the two main streams of ideas that came together to form NPM. The first, and in contrast to traditional military-bureaucratic ideas of good administration that emphasized orderly hierarchies and elimination of duplication or overlap, was the new institutional economics movement. New institutional economics from early works such as Black (1958) and Arrow (1963) contributed ideas of contestability, user choice and transparency. The second contributory stream of ideas came from business-type "managerialism", in the tradition of the international scientific management movement (Hume 1981, Merkle 1980, Pollitt 1990). These ideas emphasised the value of professional management expertise as transferable and necessary to govern technical expertise, as requiring high discretionary power to achieve results and central to better organizational performance. These two streams can conflict and as a result, different countries adopted models with varying dominance of either one stream or the other (Hood 1991).

The aforementioned World Bank study showed that administrative reformation was gradual, problematic and varied greatly from country to country. In the attempt "to transform public administrations from centralized, party dominated bureaucracies into modern, efficient, performance-oriented ones" (Nunberg et al. 1999: 2) there was varying success.

Transition in the CEE region was unique. There was no historical model, set of proven principles or pre-meditated sequence to guide the rapid flow of events. Other historic transitions include the post war democratic changes that took place in Italy, Japan and West Germany, the democratic changes to countries of the Mediterranean in the 1970s (Portugal, Spain and Greece) and the collapse of authoritarian regimes in South America in the 1980s (Argentina, Brazil, Uruguay, Paraguay and Chile). These however, were transitions to democracy with capital remaining largely in the hands of its original owners. The transition from communism required reformation of the entire economy and the creation of a previously non-existing entrepreneurial class (Offe 2004).

Amidst the economic crises of the early 1990s that resulted from transition and afflicted many post-Soviet states, science and development was not at the top of policy makers' agendas (Chataway 1999). As we mentioned above, the inability of policy makers to direct every economic sector, coupled with a purposive attempt to reverse the extremely interventionalist role of the state during the socialist regime, could have led to a withdrawal of state control from many policy areas (Von Tunzelman 2005) including the life sciences. Neglect under communism, and the historical separation of public sector research from industry have been detrimental to the current life science sector, adding weight to the oft quoted adage "history matters" (See Chapter 4 by T. Kulawik, Section 4.4, this volume).

Senker et al. (Chapter 2, this volume), find that given this shared history of institutional separation between public sector research and industry, economic crisis and neglect of the life-sciences, countries of CEE have varied in their abilities to overcome the barriers to the development of a successful life science sector. In Bulgaria significant capacity existed in fermentation technology and pharmaceutical biotechnology prior to transition. During the communist regime Bulgaria produced pharmaceuticals for its domestic market and for other CEE states. Transition led to a collapse in demand from these markets, and Bulgaria was unable to enter Western markets because its products did not meet the requirements for handling genetically modified organisms. As a consequence of transition, biotechnology-related infrastructures, resources and capacities in Bulgaria deteriorated and have not shown significant recovery in comparison to the other CEE countries in their sample. The study also finds that for certain other countries in CEE (Slovenia, the Czech Republic and Hungary) there is capacity building in biotechnology and a gradual convergence towards European neighbours. One factor contributing to this progress are the policy instruments in place to support public/private research collaborations and networking.

The importance of institutional relationships between actors of innovation systems such as those mentioned above, and the influence of shifting institutional relationships have been examined by authors such as North (1990), Nelson (1992), Cooke (1997) and Etzkowitz and Leydesdorff (2000). These frameworks which examine national and regional innovation systems and the complex triple helix relationship between academia, industry and government, do not fully accommodate the dynamics contributed by downstream actors such as publics and the political environments within which they operate.

10.3 Integrating New Actors in a New Landscape

Economic transition and EU convergence has instigated the need for a more diverse range of downstream actors to be involved in discussions on the use and development of life science technologies. This volume has explored what this has meant in a context where institutional relationships have shifted in the course of transition and a more general international movement in public administration towards NPM.

Two problems emerge from the discourse on NPM. Firstly, as public administration takes on an apolitical business-type managerialism, government might distance itself from the use of participatory and deliberative tools, choosing instead to utilise "expert policy making cultures". Lyall et al. (2009) in a specific discussion of the life sciences and policy making, frame the problem in a more nuanced way; the potential incompatibility between evidenced-based decision making and greater stakeholder engagement. Second, the concepts of transparency and accountability have been central topics in conversations around good governance. They have been more or less universally endorsed without much reflection. Transparency and accountability on the one hand may be compatible principles, with accountability implying the sharing of information and transparency revealing to whom and how actors are accountable. On the other hand, transparency and accountability may form a less comfortable relationship (Hood 2010). For example, full transparency might lead to low–intelligence defensive, box-ticking, one-way communication, rather than effective dialogue (O'Neill 2002, 2006).

The points of unease between transparency and accountability surface in the accounts of the life science sector in CEE. Griessler (Chapter 3, this volume) describes a legitimacy deficit in biotechnology and life sciences. Regulation and policy making in the EU with regards to biotechnology and the life sciences is technocratic and science driven. This prevails, despite an emerging rhetoric of participation; "transparency, accountability and participatory approaches in public policy-making need to be reinforced" (Commission of the European Communities 2001). In the case of Human Embryonic Stem Cell (HESC) research, Griessler follows the decision process of the European Commission, and analyses the use of the expert advisory body, the European Group on Ethics and New Technologies. Commission policy on the funding of HESC within the framework programmes was based on the recommendations of this expert advisory body. The MEPs of the European Parliament on the other hand "act as a conduit for the expression of a diverse range of ethical views". However, the decisions on HESC were eventually taken by a small number of top politicians. The European Group on Life Sciences was allocated the task of engaging an informed public. They attempted this through a conference in 2001, attended by approximately 800 people. The results of the conference were vague, though the EGLS group formed quite definite conclusions that did not necessarily reflect the deliberations that took place. Griessler concludes that HESC research continues to be an elite concern in the EU. Accountability in this story was achieved by the use of strict administrative routines by the commission, by the utilisation of expert opinions and inclusion of bioethics reviews. The decision making process through these means is perhaps less transparent. Bryant (2009) describes the EU as a rule-driven bureaucracy. Legitimacy in EU policy making processes is derived from the relationship between the Commission and the European Parliament and a highly formalised consultation process. Bryant similarly argues that the ability to identify the legitimacy of groupings coming together outside of this regulated process, is restrained. The argument that a group gains legitimacy from the process that it has followed is difficult to understand by bodies who are wedded to notions of representative democracy.

Kulawik in Chapter 4 of this volume discusses the HESC in Poland. At the time of accession, an attempt was made to initiate an informed discussion with various publics and groups. This included a conference, round table discussions, a public lecture and an internet forum. At the time of writing, the analysis of the internet forum, and its publication in the public domain, as promised by the Polish government, had not been completed. Kulawik reports that although consensus had not been reached, the government's statement expressing its approval of the use of embryonic stem cells under certain conditions, remained unchanged. Kulawik highlights the relationship between the Polish government ministries and their expert advisory committees as lacking transparency. The workings, form and quality of the cooperation between the consultants and ministries are largely unknown. "In the Sejm and Senate the chairs of the Committees are responsible for consulting external experts within the legislation process. To my knowledge there are no analyses of this practice" (Chapter 4 by T. Kulawik, this volume).

Krajewska (Chapter 5, this volume), writing on the legal ambiguities of medical genetics in Poland adds that regulators, being pre-occupied with democratisation and liberalisation, have not introduced specific regulations with regard to medical genetics. And also both law and society have failed to keep pace with the shift from genetics to genomics and from genomics to post genomics. We might note the additional challenges faced by economies in transition. While the much larger and more experienced EU institutions have difficulties in balancing transparency and accountability, the smaller CEE administrations must do this whilst recovering from economic crisis and trialing new methods of public engagement in discussions on new technologies. In the gap between the introduction of a new technology and an informed public, the gap marked as "catch-up", there is a reliance on experts, for example, the bioethics committees mentioned by Griessler.

The cases outlined in Chapters 3, 4 and 5 in particular, demonstrate some of the problems with the involvement of expert groups in decision making. Often there is a perceived obscuring of transparency. The process alone is not enough to communicate accountability and legitimacy to the wider public, questions of representation are raised. We can see that the problems are not exclusive to CEE and are well documented in other sources. They include for example the possibility that the privileging of expert knowledge restricts forms of argumentation and the issues available for deliberation (Bryant 2009).

Collins and Evans (2002) discuss in detail the role of experts and expertise in science. Science studies has experienced three "waves". The first wave sought to understand why science has been so successful. It took science to be esoteric with a top-down flow of information from scientists who had special access to the truth. There was no question of legitimacy. With the work of Kuhn in the 1960s, this first wave gave way to a second alternative view of science. The second wave, also called social constructivism, reconceptualised science as a social activity. Advocates conclude that decision making in and around science can and should be widened beyond the core of certified experts. How far beyond the core is a question introduced by Collins and Evans in a third wave. These authors categorise expertise and examine the role that experts play in decision making in different types of science. A central problem these authors identify is that decisions of public concern have to be made according to a timetable established within the political sphere, not the scientific or technical sphere. Because the pace of politics is faster than the pace of scientific consensus formation, decisions need to be made before the "scientific dust has settled", and before scientific consensus has been reached.

Wave one assumed a clear distinction between certified experts and the lay public, with a top down structure of information flow. Wave two deconstructed these boundaries, with the scientific community seen as indistinguishable from the citizenry as was the distinction between scientific expertise and political rights. Wave three argues that there are pockets of expertise within the citizenry and within the scientific community. This expertise is not of the certified kind, but is grounded in experience. Collins and Evans therefore call for a reconstruction of boundaries and a re-categorisation of expertise into the following three levels;

(1) No Expertise: That is the degree of expertise with which the fieldworker sets out; it is insufficient to conduct a sociological analysis or do quasi-participatory fieldwork.
(2) Interactional Expertise: This means enough expertise to interact interestingly with participants and carry out a sociological analysis.
(3) Contributory Expertise: This means enough expertise to contribute to the science of the field being analysed.

It follows then that as experts come together, to contribute to, or affect change in society regarding a new scientific development, there should be interactional expertise in each expert community to ensure the combination and translation of knowledge is done with integrity, particularly where the interests of the wider community are to be represented.

Many of these themes resound in Chapters 7 and 8. Chapter 7 begins by describing an uncertain political climate stemming from the lack of a coherent transition strategy and unsystematic privatization in Hungary (Spicer et al. 2000). Uncertainty has led to an undermining of civic trust in public institutions, hindering the development of a "civil society". In place of the trust, networks and civil society that Fukuyama argues is essential to the development of social capital in a society (Fukuyama 2001), is a "Partocracy" (Ágh 2001). Political parties and party politics dominate discussions on most issues, including biotechnologies and GMOs (Chapter 7). There is little voice or activism from national NGOs representing consumer groups or environment groups. Here we can see the managerialism and the expert-basis of decision making as described by Hood. These do not arise through a gradual evolution of the political and administrative system, but have been instituted by default in a society where rapid change during transition, with the formation of new elites, increasing poverty, social prejudices and integration of minorities, proved to be a much more acute problem than biotechnology (Chapter 7). The frames of public discourse are further defined by other dominant actors which arguably through their lobbying power, are amongst those political elites participating in policy discussion; biotechnology companies and academic biotechnology research institutions. The platforms, arenas and forums where public participation traditionally contribute to participative or deliberative democracy, seem to be lacking in this context, as does the translational knowledge and experience needed to link communities and translate between them (Collins and Evans 2002).

Chapter 8 similarly analyses the consequences of policy uncertainty for a particular community in Hungary. Uncertainty in this case undermines future planning and investment in biotechnology. The communities studied are the same academic biotechnology research institutions that feature in Chapter 7. Where Chapter 7 evaluates this group and its role in the policy discussion through the published and dominant opinions on the use and value of GM, Chapter 8 asks different questions of this group of scientists and finds that they are deeply affected by uncertainty in government policy. The chapter illustrates that while the group may be perceived as being one of the policy elite contributing to policy discussion, there remains a disconnect between government policy makers and this group of upstream users and developers of the technology. The interactional expertise and knowledge translation that the third wave describes as necessary again is detrimentally absent.

Chapter 6 describes the population genome project and organ transplantation in Latvia. The important role of interactional expertise becomes evident in the discussion of risk and trust which surrounds the two biotechnologies. In the case of the genome project the public amongst other things are concerned with power and control over genetic information, misuse of data and anonymisation. Putnina writes that the media serves as a mediator between science and society. So too could interactional experts. Putnina's discussion of the two cases also challenges Evans and Collins' view somewhat. The adoption of a technology by society seems more than a problem of knowledge translation between experts and the public through interactional expertise. Chapter 6 suggests that attitudes towards biotechnologies are not stable, but dependent upon past experience. Technology users and the public build reflexive accounts of technologies. If end users are not integrated into the planning and discussion at the earliest stages of development and adoption, this then may pose problems later down the line. The public becomes a legitimate provider of input just by virtue of being the end user. Technology policy has a significant contribution to make here.

Chapter 9 moves us finally to consider the concept of space. Harmonisation in Europe has increasingly meant the creation of an economic space that facilitates the free movement of goods, services, labour and capital. This homogenised and abstract space requires that social relations are progressively detached from their situatedness. GMOs had to be similarly stripped of their cultural meaning in order to move around this uniform EU regulatory space. Reynolds and Szerszynski demonstrate how the abstraction of space poses problems for technologies and societies which are mutually constitutive (Mackenzie and Wajcman 1999). They find that GM crops do not stand separately from society, that this technology has been shaped by social relations and vice versa.

The discussion so far has highlighted the following four findings that emerge from this volume.

1. It is important that CEE policymakers incorporate different actors into the policy making process, in order to widen expertise and ensure legitimacy, accountability and transparency.
2. It is impossible to disassociate technologies from social relations.
3. There are tensions and dynamics between EU regionalization and the national agendas of different CEE countries. The EU regulatory space may indeed have been an attempted technocratic solution to create an abstract space for the free movement of technologies, but we have found that technologies are first debated in national spaces. National spaces are differently constituted.
4. The role of actors, including experts and those representing civil society groups, varies from country to country, as does the influence of each actor and route to legitimacy. Each case is unique, and bound up in the cultural traditions and history that were briefly described above.

Reynolds and Szerszynski find that the attempt to create a standardised regulatory space resulted in tension between the neo-liberal approach advocated by the OECD and the precautionary instincts of some member countries. The cultural diversity within the EU and local agricultural visions caused the breakdown of the homogenous EU regulatory space back into national units.

10.4 The Contours of the CEE Biotechnology Landscape

In his cartographic representation of cultural boundaries of science and technology, Gieryn (1999) drawing from earlier work by Foucault and Bourdieu wrote of knowledge-making in terms of space, similar to Reynolds and Szerszynski, using the analogy of map creation. He used it to consider what is included, what is excluded, what is the physicality of the knowledge terrain; the mountains, hills and streams, capitals and small towns. More recently, this physicality, and the complexity with which different actors and institutions weave a different pattern for each biotechnology case, has been referred to in textile nomenclature as the "texture" of the biotechnology "fabric" (Bruce 2011). We find these physical analogies useful in helping to understand what is unique about biotechnology in Central and Eastern Europe. Having started with our earlier special issue on biotechnology innovation in CEE, and focusing on the triple helix (Etzkowitz and Leydesdorff, 2000) of the relationship between the academy, state and economy in producing biotechnology knowledge, we have now added further layers to our map of the landscape with a focus on civil society actors, public engagement, and biotechnology governance, layers that are often missing in triple helix approaches. This creates a more complex and dynamic picture of the innovation context. In our review of the chapters of this volume, we find that there is a common set of themes, or contours, emerging in each biotechnology case. These include:

1. The roles of key institutions including the politics of expertise,
2. Modes of public engagement, including whether and how publics are included in decisions about biotechnology, and
3. Policy outcomes that emerge both from the communist legacy and EU accession.

We discuss these in Tables 10.1 and 10.2 below.

Table 10.1 is a snapshot of a second layer of findings that emerge from this volume. In particular, two points become apparent (see Table 10.1).

1. The role played by the EU harmonization of national institutions is important.

In a number of cases, authors found both the harmonization and transition processes to be problematic, in particular those who looked at agricultural biotechnology. While EU accession provided new markets and funding sources, this was outweighed by the weakened state of the scientific institution following transition, and challenges of how to best serve and protect CEE small, medium and large agricultural producers in a European policy context. Those who looked at human biotechnology tended to see more opportunities in transition and EU accession, with the exception of Greissler who found it facilitated an expert-led approach to decision making. The Polish case is a compelling example of the positive outcome of the liberalization and democratization agenda, which provided opportunities to create new policies that more flexibly adapted to a post-genomic era. Here Catholicism and Catholic belief systems were wrapped up in the Solidarity struggle against an atheist communist regime, as such morally-driven notions such as "the conceived child" replaced the more scientific socialist concept of the "embryo".

2. Public engagement is at an early stage of development in Central and Eastern Europe.

Table 10.1 Contours of the biotechnology landscape

Chapter and condensed title	Institutions	Public engagement	Policy outcomes
2. Biotech in CEE	EU harmonization of key institutions	Expert-led	Central planning and transition weakened science
3. Citizen Participation in EU Stem Cell Research Policy	EU harmonization of key institutions	Rhetoric of participation, but not delivered	Europe leads on public engagement, but it is still expert-led
4. Politics of Embryo Research in Poland	Religious institutions shape biotech policies on embryo research	Expert-led	Liberalisation and democratization are more important than S&T, ironically this can bring flexibility in some biotech policies
5. Legal Ambiguities of Medical Genetics in Poland	Religious institutions shape biotech policies on embryo research	Expert-led	Liberalisation and democratization are more important than S&T, ironically this can bring flexibility in some biotech policies
6. Managing Risk and Trust in Latvian Human Biotechnologies	Public trust in scientific and medical institutions is high	Expert-led, but there is an opportunity for patient groups to be involved	Post communist era brings opportunities for building trust
7. GM food Battlefield in Hungary	NGOs, industry and media frame public debate	Publics are represented by NGOs and consumer groups	EU accession provides new markets, but also dilemmas on how to protect farmers
8. Agricultural Biotech Research in Hungary	Tension between EU harmonization and national institutional agendas	Consumers shape policy, but publics are not involved in decision making	EU accession opens new funding, but the transition generally undermined science
9. GMOs and European Agro-technological futures in Poland	Tension between EU harmonization and national institutional agendas	Social movements & small farmers resist harmonization	EU accession undermines rural agro-social structure of CEE

Table 10.2 Detailed Analysis of the CEE Biotechnology Landscape

2 Biotechnology in Central and Eastern Europe: an Overview		
Institutions	Public Engagement	Policy Outcomes
• Funding instruments: Biotech-specific policy instruments, generic policy instruments, EC's Sixth Framework Programme • Research instruments that encourage networking • S&T policy actors e.g. ministries for economy, or education and research. • Advisory bodies to ministries (include representatives from academia, industry and funding agencies).	• Experts who sit on advisory bodies reporting to ministries represent publics.	• Central planning weakened the life sciences in many post communist countries. Historic separation between academic research and industry. • Transition caused deterioration of capabilities, resources and infrastructure related to life sciences.
3 Citizen Participation in Controversial EU Research Policies? The Debate on Human Embryonic Stem Cell Research within the 6th Framework Programme		
• European Commission and its scientific advisory committees (e.g.ELGS, EGE); role in fostering harmonisation for economic reasons and delivery of framework for ethical evaluation of HESC research. The commission follows a expert-model. • The European Council (divided into permissive and restrictive factions mirroring dominant religious beliefs, economic and strategic interests). • The European Parliament • Outside experts	• A strong rhetoric of participation which influences EC (in conflict with use of experts). However, formal procedures, powers and negotiation at top levels limited openness of debate. "Public was restricted to a spectator on the internet". • Call for scientific and ethical evaluation of FP5 research projects involving HESCs to be made public and transparent. • Civil society interests are expressed through European Parliament electives. • European ethics principles formed by expert advisory committees involve limited consultation with established stakeholders. • Directly and indirectly informing publics	• Some European institutional innovations in policy making to include publics in discussions on ethics, e.g. workshops, seminars, meetings, conferences. However, these still follow a traditional expert-led model.

Table 10.2 (continued)

4 The Politics of Human Embryo Research in Poland		
Institutions	Public Engagement	Policy Outcomes
• The church has played a role in shaping the biotechnology landscape, but the church alone is too simple an explanation for a complex process. • The church's conservative attempts to restrict abortion were supported by the Solidarity Movement since at least 1989. • A key moment came in the early 1990s with the shift in language from medical terms "embryo", "pregnant woman", "doctor" and "hospital" to the value-laden "conceived child", "mother", "aborter", and "abortion chamber". • Poland was to be an example of 'moral order' since it couldn't be one of "economic advancement" - rooted in a narrative of the heroic history of Poland.	• The abortion struggle has shaped the public sphere, creating a hegemonic paradigm of "public morals" that influences policy oriented discussions of bioethical dilemmas. • Poland does not as yet have the institutional capacity for public discussion of these issues. • What has occurred has been a "quasi-public" consultation between the rivals in the abortion war' (including scientists, ethicists, and church representatives) on stem cells, but this was not genuine public engagement. • The abortion public debate often slips unhelpfully into discussions of eugenics. • Opinion-formers manipulate the public through language use, stifling authentic debate.	• From 1956 to 1993 Poland had a very liberal abortion law. • Similar to other post communist states, abortion was an important part of the political agenda during the transition phase. • Even though the embryo has been defined since 1993 as a "conceived child", with strong moral connotations, in Polish legislation, there are few regulations concerning assisted reproduction and gene technology. • In contrast to theorists such as Jasanoff who see S&T issues as crucial to deliberative democracy, GMOs have not attracted the same attention in Poland as issues to do with liberalisation and democratisation.

Table 10.2 (continued)

5 Legal Ambiguities Concerning Medical Genetics in Poland		
Institutions	Public Engagement	Policy Outcomes
• Church shapes views and policy towards the moral and legal status of pre-born human life. • The church disapproves of some aspects of the liberal agenda; • Notably being instrumental in government policies that curtail research on the human foetus and embryo.	• New methods of public engagement • Elites must create effective and transparent mechanisms for public debate, which will allow for informed decisions. • Limited public involvement in decision-making so far.	• Strong liberalisation and democratisation agenda • Polish regulations, governance and its constitution are advanced and flexible in many respects, surpassing the EU, especially with regard to individual consent, and protection of sensitive information on individuals. • Overall, well-suited in many ways to the post-genomic agenda.
6 Managing trust and risk in new biotechnologies: the population genome project and organ transplantation in Latvia		
• At an institutional level, there is a disconnect between organ/gene donation as gift and as commodity, which can heighten risk perception and erode trust.	• Publics are sceptical of optimistic framings on transplantation, but in Latvia people remain highly trusting of scientists (95% trust scientists). • Patient groups have not, as yet, shaped the biotechnology debate in Latvia, but they have great potential to do so.	• Regulation should focus on institutionalising reciprocity and trust, which is based most strongly on one-to-one relationships, such as between doctor and patient, or gene donor and scientist. • As yet in gene donation, regulation focuses on making moral relations abstract, which heightens risk perception and erodes trust. • In contrast, heightened transparency in organ transplantation enhances trust, since it enables face-to- face communication between donor and recipient.

Table 10.2 (continued)

7 Social Trenches in the GM Food Battlefield: Experiences of a Survey Series in Hungary		
Institutions	Public Engagement	Policy Outcomes
• Publics; trust and risk perception plays a major role in the acceptance of new technologies. • NGOs (representing for example environmental concerns and consumer issues) articulate concerns/criticisms. • Political parties; "partocracy" causes the politicization of technology issues. • Industry and industry associations; active in lobbying government. • Agricultural producers; lobby government and promote technologies. • Mass media plays an important role in opinion forming. • Government ministries, departments and regulatory agencies. • Academic institutions (e.g Academies of Sciences). Academia is divided and polarized on GM crop issues.	• Representation of some publics through NGOs and consumer groups which protest, lobby government and publish information. These represent a less sophisticated and active communication strategy than other actors in Hungary also involved in debates on GMOs.	• Accession into the EU has widened the market for agricultural produce. Policy making faces a dilemma in whether to support models for agriculture better suited to small and medium sized family run farms or larger farms that can compete with farms in Western Europe. Hungarian agriculture has not been able to develop a coherent, widely accepted agricultural strategy and the debate on GMOs is ongoing despite European regulation.

Table 10.2 (continued)

8 Coping Strategies and System Adaptation of Agricultural Biotechnology Research in Hungary		
Institutions	Public Engagement	Policy Outcomes
• European regulators encourage harmonisation through the framework program funding. • National regulators and government ministries are hesitant to set policy on GM crops, leading to uncertainty in R&D. • National and international markets for agricultural products. • Public research institutions (e.g. academies of sciences, universities, applied research institutes). • Industry (including small and medium Hungarian biotech firms and large multi-nationals). • National funding bodies (e.g. OTKA, NKTH) play an important role in allocating funding to the development of crop biotechnology. There are issues around transparency and the reasons as to why funding is allocated and strategies for science and R&D are not clear.	• Public voice and consumer choice as expressed through the market is a very powerful signal for national policy makers and funders of scientific research. • Currently there is a noted lack of more direct public involvement in policy making in Hungary possibly due to the non existence of fora or arenas.	• Transition and fiscal austerity has resulted in a lack of funding and uncertainty for the science system. • EU accession has opened up opportunities for scientists to access EU funding sources and building partnerships.

Table 10.2 (continued)

9 Contested agro-technological futures: the GMO and the construction of European space		
Institutions	Public Engagement	Policy Outcomes
• At an institutional level the European political, economic and cultural project has sought to harmonise, or standardise, differences between countries, potentially losing individual difference; it is strongly based in a neo-liberal paradigm. • The GMO itself is a homogenising technology.	• Attempts at homogenisation perhaps ironically attract response by social movement organisations and publics resulting in heterogeneous national responses, as in the conflict over GM crops. • This was particularly the case in Poland, which is based around small family farms inconsistent with the GM agro-industrial paradigm, sparking the creation of many 'GM free zones' in the country, and strong consideration of organic agriculture as a main economic activity.	• GMO agriculture is based on an agro-industrial paradigm, requiring a homogenised European regulatory space. • This is in contrast to an ecologically-integrated paradigm, which seeks to understand complexity of agro-ecological relationships. • The rural agro-social structure of nations such as Poland should not be viewed as pre-industrial or the product of communist legacy, but as the basis for a post-industrial agrofood future. • They could also form a possible path to an alternative ecological modernity.

In many cases it was expert-led by committees representing the public, or by stakeholder organizations representing different public interests, including consumer groups. Many wrote of the missed opportunity for public engagement, where there are future possibilities for integrating key public groups such as patients in the case of Latvian health biotechnologies, but in general the post-communist era provided opportunities for building trust.

Table 10.2 provides a more detailed comparative analysis of each chapter, providing a third layer of findings (see Table 10.2).

1. There are clear harmonization efforts at the EU level.

For GM crop technologies, this is more apparent than for HESC research for example. Crops require uniform standards for production, safety and labeling in order to allow the internal EU market to function as intended. However, this results in a tension where national governments attempt to set more or less stringent standards according to their individual priorities, ideals and goals. The analysis of key actors shows how and through whom these national differences are expressed and shaped. Dominant actors in national settings tend to be those traditionally overlooked by economic and innovation systems theories, but are nevertheless important social actors which play a role in communicating to and for publics. These include the Church and the media.

2. There is a general lack of public engagement and inadequate public representation.

Countries in CEE are new democracies where public engagement is relatively unpracticed. This means there is an opportunity to take advantage of a 'clean-slate' and build the institutions for democratic governance of new technologies from the ground up. It seems difficult to learn any lesson from Western Europe, because successful governance and participatory exercises are rare even in more developed parts of the region. At the same time, in CEE challenges arise from institutional infrastructural weakness resulting from the pre- and post-transition period and the associated economic crises.

3. Different areas of policy shape different biotechnologies to varying degrees.

Economic and trade policy has a greater impact on GM crops for example, while health technologies are shaped more by moral questions that ask about the nature and beginnings of human life itself. As a result while biotechnology in the region has common landscape contours, the role of key institutions, challenges with public engagement, the communist legacy and difficulties with transition, the patchwork tapestry of the whole region reveals a great deal of complexity, and the individual uniqueness of each national and biotechnology case.

10.5 Conclusions

This volume has shown that in a certain number of CEE countries, there are indications that there is capacity building in the life science sector and convergence with Europe. Because the life sciences are a key enabling technology, because countries seek to join the European Union and because the life sciences stand to impact the life of ordinary citizens, countries must develop legislation and policy instruments for the development and use of technologies. In this endeavour, the authors of this volume would argue that a more diverse range of actors need to be involved.

Another argument pursued in this book is that the process of transition did not involve the creation of institutions to build social capital and civil society. In other words the platforms, arenas and knowledge translation mechanisms needed to involve the public in political process and discussions on life science technologies effectively. Exercises in public deliberation and participation require scarce resources and experience.

The problems associated with engaging publics in discussions on novel technologies are experienced everywhere. Every society faces the problem of achieving representation of publics, legitimacy, accountability and transparency. However, these challenges are particularly acute in Central and Eastern Europe where many societies make use of experts to engage in debate and make decisions on behalf of their citizenry. The use of experts is especially familiar in the context of EU technocracy. However, the accounts described here, caution against the use of experts and the process of decision making where experts and other dominant actors fill the space where public engagement is simply missing. It is the challenge of the authors of this volume that CEE policymakers give special consideration to legitimacy, accountability and transparency in order to build public trust. It is through this process that Central and Eastern Europe's biotechnology landscape will achieve a sustainable socio-ecological future.

References

Ágh, A. 2001. Early democratic consolidation in Hungary and the Europeanisation of the Hungarian polity. In *Prospects for Democratic Consolidation in East Central Europe*, Pridham, Ge offrey & Ágh, At tila (eds). Manchester & New York, Manchester University Press.

Arrow, K J. 1963. *Social choice and individual values*, 2nd Edition. New York, Wiley.

Bache, I. 2003. Governing through governance: Education policy control under new labour. *Political Studies* 51(2):300–314.

Black, D. 1958. *The theory of committees and elections*. Cambridge, Cambridge University Press.

Bruce, A 2011. The Textures of Controversy: Values and Interests in Disputes around Genomics, PhD Thesis, Milton Keynes: Open University.

Bryant, P. 2009. Deliberative governance: Political fad or a vision for empowerment? In *The limits to governance. The challenge of policy-making for the new life sciences*, eds. C. Lyall, T. Papaioannou, J. Smith, 349–378. Aldershot, Ashgate.

Chataway, J. 1999. Technology transfer and the restructuring of science and technology in Central and Eastern Europe. *Technovation* 19:355–364.

Collins, H and R, Evans. 2002. The third wave of science studies: Studies of expertise and experience. *Social Studies of Science* 32(2):235–296.

Commission of the European Communities. 2001. Towards a strategic vision of life sciences and biotechnology: A consultation document. COM(2001)454 final, Brussels, 4.9.2001.

Cooke, P. 1997. Regional innovation systems: Institutional and organisational dimensions. *Research Policy* 26(4–5):475–491.

Etzkowitz, H and L, Leydesdorff. 2000. The dynamics of innovation: from National Systems and "mode 2" to a Triple Helix of university-industry-government relations. *Research Policy* 29:109–123.

Fukuyama, F. 2001. Social Capital, Civil Society and Development. *Third World Quarterly* 22(1):7–20.

Gieryn, T. 1999. *Cultural Boundaries of Science: Credibility on the Line*. London, Chicago University Press.

Hood, C. 1991. A public management for all seasons. *Public administration* 69:3–19. http://newdoc.nccu.edu.tw/teasyllabus/110041265941/Hood%20NPM%201991.pdf.

Hood, C. 1995. The new public management in the 1980s: variations on a theme. *Accounting Organisations and Society* 20(213):93–109. http://www.drmanage.com/images/1202965572/Hood_NPM(1995).pdf.

Hood, C. 2010. Accountability and transparency: Siamese twins, matching parts, awkward couple? *West European Politics* 33(5):989–1009.

Hume, L J. 1981. *Bentham and bureaucracy*. Cambridge, Cambridge University Press.

Jeffries, I. 2002. *Eastern Europe at the turn of the twenty first century. Routledge studies of societies in transition*. London, Routledge.

Lyall, C, T, Papaioannou and J, Smith. 2009. *The limits to governance. The challenge of policy-making for the new life sciences*. Aldershot, Ashgate.

Mackenzie, D and J, Wajcman. 1999. *The social shaping of technology*. Milton Keynes, Open University Press.

Merkle, J. 1980. *Management and ideology: the legacy of the international scientific management movement*. Berkeley, CA, California University Press.

Nelson, R. R. 1992. National innovation systems: A retrospective on a study. *Industrial and Corporate Change* 2:347–374.

North, D. 1990. *Institutions, institutional change and economic performance*. Cambridge: Cambridge University Press.

Nunberg, B, et al. 1999. The state after communism: administrative transitions in Central and Eastern Europe. The International Bank for Reconstruction and Development/The World Bank. Washington

Offe, C. 2004. Capitalism by democratic design? Democratic theory facing the triple transition in Central Eastern Europe. *Social Research* 71:3.

O'Neill, O. 2002. *A Question of Trust*, Cambridge, Cambridge University Press. (The 2002 Reith Lectures).

O'Neill, O. 2006. Transparency and the ethics of communication. In *Transparency: The key to better government*, eds. C. Hood and D.A. Heald, 75–90. Oxford, British Academy/Oxford University Press.

Pollitt, C. 1990. *Managerialism and the public services: the Anglo-American experience*. Oxford, Blackwell.

Radosevic, S and L, Auriol. 1999. Patterns of restructuring in research development and innovation activities in Central and Eastern European countries: An analysis based on S&T indicators. *Research Policy* 28(4):351–376.

Spicer, A, G A, McDermott and B, Kogut. 2000. Entrepreneurship and privatisation in Central Europe: the tenuous balance between destruction and creation. *Academy of management Review* 25(3):630–649.

Von Tunzelman, N. 2005. Network Realignment, Policy Sustainability and EU Accession. In *Science and Technology Lessons for CEE Countries*, Inzelt, A Zaman, G and Sandu, S. Bucharest, Expert Publishing House.

Index

P.T. Robbins, F. Huzair (eds.), *Exploring Central and Eastern Europe's Biotechnology Landscape*, The International Library of Ethics, Law and Technology 9,
DOI 10.1007/978-90-481-9784-2,